THE RADIO PHON

Justin Patrick Moore started his radio career in 1999 on Anti-Watt, a pirate station at Antioch College, Ohio. Between 2001 and 2014 he was one of the rotating hosts for the experimental music shows Art Damage and On the Way to the Peak of Normal on WAIF, Cincinnati. In 2015 he became a ham radio operator with the call sign KE8COY and started making friends in the shortwave listening community. These connections led him to contribute regular segments for the high-frequency programs Free Radio Skybird and Imaginary Stations. He first started writing about music for Brainwashed.com and his essays and short fiction have appeared in a variety of venues.

First published by Velocity Press 2024

velocitypress.uk

Printed and bound in Great Britain by Clays Ltd, Elcograf S.p.A.

Cover design

Hayden Russell

Typesetting

Paul Baillie-Lane

www.pblpublishing.co.uk

Line illustrations © Aisha Nuraini/Dreamstime

Cover image credits – from top to bottom

1. Hieu Nguyen

2. Mister rf: https://creativecommons.org/licenses/by-sa/4.0/deed.en

3. no attribution required

4. Stefan Schweihofer

5. Internet Archive Book Images

6. rawpixel.com / U.S. Department of Energy (Source)

ISBN: 9781913231552

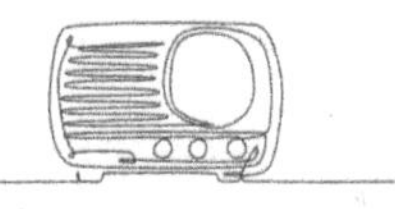

The Radio Phonics Laboratory

Telecommunications, Speech Synthesis and The Birth of Electronic Music

Justin Patrick Moore

Contents

This book is dedicated to:

my dear wife Audrey L. Cobb
with whom I like to ride on a bicycle built for two

and

my father Victor Justin Moore, who put up with a lot of noise when I was a teenager and who has helped me out with spiritual and practical wisdom, handyman know-how, and backyard mechanics as I pursued my callings.

First Utterance

In Francis Bacon's unfinished utopian novel, *New Atlantis*, posthumously published in 1626, the British natural philosopher and "father of the scientific method" wrote of a place that had yet to exist, where scientists and sound researchers explored the nature of acoustics through the use of a new class of instruments, inventing never-before-heard sounds while conveying them across great distances. He wrote:

> *We have also sound-houses, where we practice and demonstrate all sounds and their generation. We have harmonies, which you have not, of quarter-sounds and lesser slides of sounds. Divers instruments of music likewise to you unknown, some sweeter than any you have, together with bells and rings that are dainty and sweet. We represent small sounds as great and deep, likewise great sounds extenuate and sharp; we make divers tremblings and warblings of sounds, which in their original are entire. We represent and imitate all articulate sounds and letters, and the voices and notes of beasts and birds. We have certain helps which set to the ear do further the hearing greatly. We also have divers strange and artificial echoes, reflecting the voice many times, and as it were tossing it, and some that give back the voice louder than it came, some shriller and some deeper; yea, some rendering the voice differing in the letters or articulate sound from that*

they receive. We have also means to convey sounds in trunks and pipes, in strange lines and distances.

It would be more than two centuries before Bacon's so-called "sound houses" began to manifest.

The history of electronic music is inseparable from the history of radio and telecommunications. Beginning in the nineteenth century, electricity opened up new worlds of sound, from the dits and dahs of Morse Code, to the crash of static and detuned voices heard over radio. Time and time again, the same breakthroughs and devices invented by electrical engineers for communicating in telegraphy, telephone, and radio were quickly adapted for use by musicians. Elisha Gray, one of the inventors of the telephone, heard music in the oscillating sounds of the telegraph, leading to his invention of one of the first electronic music instruments, the Electro-Harmonic Telegraph. Gray's work prefigured the way oscillators would later be used as the basic building block for electronic sounds. Thadeus Cahill coined the word synthesizer and invented the first one that he intended to use to transmit music over the telephone. Lev Theremin's work as a radio engineer led to him using antennas in his namesake instrument.

This book explores the ways the telecommunications field gave birth to electronic music as we know it today. We'll discover how research into the transmission, compression, and synthesis of speech over telephone lines and radio waves gave birth to the techniques and tools now used as part of the standard kit in electronic music in particular, and music production in general. In this way, electronic music can be seen as the unintended, if happy, byproduct into research on the nature of speech and its electrical transformations.

As the twentieth century got underway, musical experiments by electrical engineers and radio engineers became increasingly common. They sought to transmute electrical energy into sound and a new category of electronic instruments emerged out of their curiosity. Recording technology in the form of wire recorders, phonographs, and magnetic tape later enabled them to play with sound, rearrange it, and investigate its properties in ways that had not been done before. Halim El-Dabh used wire recorders borrowed from Radio Cairo to creatively remix the sounds of a traditional Egyptian exorcism into music. Pierre Schaeffer started off as telecommunications engineer and quickly perceived the way in which the radio test equipment could be used to create audio laboratories where sound could be manipulated and explored. He invented the musique concrète style and established the GRM electro-acoustic music in conjunction with French national broadcaster Radiodiffusion-Télévision Française.

Schaeffer's work created the template for the electronic music studios that followed in the way that they were allied to broadcast radio and television. The Electronic Music Studio where Karlheinz Stockhausen composed his groundbreaking experiments was made possible by Westdeutscher Rundfunk radio station in Cologne, Germany. The Nippon Hōsō Kyōkai broadcasting service in Japan followed suit to create their own electronic music studios, as did the BBC's mythic Radiophonic Workshop. These stations and their studios became the original innovators of the genre, who performed their experiments and made a new kind of music that was perfect for transmission over the aether.

Meanwhile, Francis Bacon's dreams were most fully realized in Murray Hill, New Jersey, where Bell Telephone Laboratories had their headquarters. They became an inadvertent laboratory of music as their research into acoustics, phonetics, speech, noise, and information theory put their scientists on the path of writing the first programs to enable computers to synthesize music.

This book is divided into three parts. The first part, *Telemusik,* concerns the invention of instruments out of the telecommunications technology that brings the distant up close: from the beginnings of the telegraph and telephone, to the radio-frequency based sounds emitted by early high voltage arc lamps, sounds made by vacuum tubes and on to the radio engineering work that influenced Lev Theremin's musical inventions. The experimental musical antics of George Antheil, his work with Hedy Lammar, and the subsequent invention of spread spectrum radio—the principle on which wifi networks are based—show how developments in the arts can influence engineering.

The second part of the book, *The Synthesis of Speech*, delves into the quest to compress speech so that more conversations could fit over the telephone wires, resulting in the invention of the voder and vocoder, which were first used in wartime cryptography before their musical characteristics were exploited. Meanwhile, the quest for fidelity, and the idea of making a machine that could listen to voiced instructions, as well as talk back, resulted in research that eventually led to the creation of the first singing computer. In time, vocoders and other principles around speech synthesis, such as linear predictive coding, led to innovation in avant-garde music circles and popular music alike.

The third section of the book, *We Also Have Sound-Houses,* looks at the creation of the first electronic music studios in association with radio broadcasters and telecommunication companies: from the wire recorder experiments first done by Halim El-Dabh with equipment loaned from Radio Cairo, to the studios associated with Radio France, the BBC, and Columbia-Princeton in collaboration with the RCA Corporation, and the experiments at Bell Laboratories, Stanford, and IRCAM that gave birth to FM synthesis and the first computer programs written to mimic the human singing voice.

In its earliest years, electronic music was a distorted maze of vibrations. Madcap inventors attempted to trace their way through the lines of circuitry to create sounds out of the chaos. Later, researchers tasked with solving a specific problem often found themselves opening the door to a vast imaginary landscape they wouldn't have even knocked on if the job hadn't landed on their desk as assigned by a manager. Later still, the avant-garde sensibilities of artists on the fringes found a foothold in the studios where they were more likely to hang out with engineers than other musicians. The conductor Herbert Weiskopf was prescient when he wrote in 1926 of the "two opposing professions, which previously had no relation with each other, namely the electrical -particularly the radio- engineer and musician now engage with one another. Out of this marriage the 'music engineer' is born, which is seen as a paradox today, but tomorrow might be taken for granted." He was right. The music engineer is now essential for everything, from live sound at clubs and concert halls, to the studios where records, television, film and video games are made.

This is the story of how the musician and the engineer joined forces to create the electronic musicians and music engineers who have made themselves at home in the sound houses of Francis Bacon. It is the story of the telecommunications scientists and idiosyncratic composers who helped give birth to the wide variety of electronic music which continues to capture the imagination of dedicated listeners around the world.

PART I

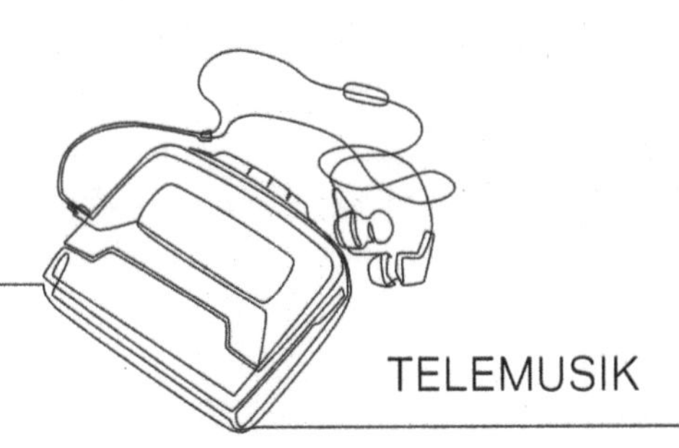

TELEMUSIK

1: Telegraphies

Visible Speech

Speech can be made visible as well as it can be made audible. The written word is the simulacrum of speech rendered to the sense of sight. Audible speech can also be written with a stylus, transcribed to wax cylinders, as Thomas Edison showed the world when he invented the phonograph.

The father of "the father of telephony," Scottish linguist Alexander Melville Bell was a man who worked hard to create a system of visible speech to be used by the deaf and hard of hearing. Thus his more famous son, Alexander Graham Bell, grew up in a household where phonetics, articulation, elocution, and linguistics abounded in the playground of the imagination. It is no wonder he went on to invent the telephone by developing a way to transcribe speech into electrical currents and make it audible again on the other side of a stretched wire.

This transliteration of words from one mode to another can be seen as part of the quest for a universal language that has been ongoing for over two millennia. Mystics, theologians, and philosophers had long been convinced that there used to be a universal language that enabled clear communication between people on all manner of topics, including the most erudite and complex. Some of these thinkers sought to rediscover a lost universal language, the original tongue people spoke before the fall of the tower of Babel, while others aimed to create one of their own. The invention of these languages was much in vogue in the nineteenth century, a game for intellectuals whose minds bent towards the way of words. As a linguist and teacher of speech, Alexander Melville Bell knew of these trends and devoted a portion of his life towards developing his own system.

These interests in translating speech to electricity and back again can be seen in the work of his son, and later in the field of telephony as people worked to encode voices and develop machines that could recognize speech. This same impulse also led to the eventual development of machines that could synthesize speech and the creation of circuits that mastered the art of utterance.

Though overshadowed by his son's frontier of interconnected wires and exchanges, the senior Bell's creation of "Visible Speech" represents the first universal phonetic alphabet, and Melville Bell designed it so that it could depict the sounds from many different languages, as well as common sounds such as coughing and sneezing.

Melville Bell was born in Edinburgh, Scotland, in 1819, to a father also named Alexander Bell. The original Alexander Bell was an expert on phonetics and speech disorders, and passed on his knowledge to his son, who studied under him and became his principal assistant. His work as an elocutionist took him from being a college lecturer in his hometown to Kingston, Ontario, and on to Boston, Massachusetts. His work centered on teaching college students, and providing elocution lessons to public speakers, deaf speakers, and those who had articulation problems such as stuttering. Today he would be known as a speech therapist.

In 1850, Melville Bell began working in earnest on his universal system of phonetics. The physiological aspect of speech was an area he chose to specialize in, and it became a key factor in the characters he created to represent sounds, making them iconic representations of the speaker's physical actions. Those who had worked on phonetic alphabets before him focused on place, manner, and voicing, and Melville Bell kept with this tradition.

The Bells were principally concerned with improving the state of affairs for the deaf, but Visible Speech could also be used by anyone who learned it to render all the languages of the world with the same visual characters. This was part of the appeal of a universal language: to solve the inconvenience presented by the towering diversity of tongues. The construction of a universal language presents enormous challenges and difficulties, however, and it was seventeen years before Melville Bell published the book *Visible Speech: The Science of Universal Alphabetics* in 1867.

Due to the fact that Melville Bell's wife (and Graham Bell's mother) was hard of hearing, forced to use an ear-tube for one-on-one conversations, the Bells already knew American sign language and the British two-handed alphabet. They would have also been familiar with Braille and Morse Code as alternate communication systems. Philosopher Marshall McLuhan pointed out in *Understanding Media* how Braille had started off as a tactile system for reading coded military messages in the dark, writing, "Letters had been codified as dots for fingers long before the Morse Code was developed for telegraph use." In their continued search for ways to help the hard of hearing, the Bells were led to look into the new field of electrical devices.

Prior to developing the telephone, one of Graham Bell's inventions was a Visible Speech Machine based on his father's work. Also called the ear phonautograph, it was based on the first phonautograph, or sound-writer, created by Édouard-Léon Scott de Martinville in 1857, which attempted to reproduce graphic traces of sound waves and was used to further the scientific study of sound. Scott de Martinville experimented with a number of natural materials to try and replicate the tympanic membrane of the human ear drum and ossicle bones.

Graham Bell and his partner in this project, Clarence J. Blake, took the next logical step and used an actual excised human ear as the mouthpiece for their second generation phonautograph. A stylus was attached to the ear drum's ossicle bone and the spoken vibrations caused it to move across a plate of glass that had been smoked with soot from an oil lamp. The stylus etched the vibrations across the sooted glass, and a trained person could read the shape of the pattern to determine what was said. The Bells felt that this Visible Speech Machine could be used to teach deaf students what vowels looked like. Though this did not work in practice, the work he had done with tympanic membrane and the ossicle bone in this invention gave him the technical skills he needed to accomplish his next feat: the invention of the telephone.

In recalling how the earlier invention influenced the other, Bell wrote,

> *I was struck with the remarkable disproportion in weight between the membrane and the bones that were vibrated by it. It occurred to me that if a membrane as thin as tissue paper could control the vibration of bones that were, compared to it, of immense size and weight, why should not a larger and thicker membrane be able to vibrate a piece of iron in front of an electro-magnet.*

The word "telephone" was around before Alexander Graham Bell was even born. According to McLuhan, it originated in 1840 and first referred to "a device made to convey musical notes through wooden rods." The word literally means "sound from afar" and it seems no coincidence that a family of people who dedicated their lives to helping the hard of hearing communicate gave birth to a device that is an electrical extension of the ear.

Four years after creating the telephone, in 1880, Graham Bell was given the Volta Prize by the French government and awarded 50,000 francs. He used this money to start the Volta Laboratory, which later became Bell Telephone Laboratories after his death. Its mission was to research and analyze sound, with a focus on its electrical transmission.

Graham Bell was like a distant father to the Volta Lab and used the prodigious funds earned by the lab for the "diffusion of knowledge relating to the deaf" and improving their lives.

The telephonic researchers at Volta/Bell Laboratories would continue on, seeking new ways to encode information over wires. As a premier facility devoted to pure research, it developed information theory, the transistor, radio astronomy, and the first musical computers. At Bell Labs they made talking machines and machines that could recognize human speech. From an artificial larynx, to the voder and the vocoder used in WWII crypto communications, to the first text-to-speech computer programs written on gargantuan mainframes, Bell Labs innovated the electronic blips and bleeps we now take for granted.

The Musical Telegraph

The stories of telecommunication and electronic music both begin with the telegraph, parent to both telephone and radio. Well before the internet, the telegraph changed the face of communication and commerce across the globe. It was an invention that revolutionized the transmission of information just as the printing press did nearly four centuries earlier.

Elisha Gray was born in Barnesville, Ohio, in 1835, the year Samuel Morse invented his system of dots and dashes, and two years before he received his patent for his electromagnetic telegraph. That very same year, Sir William Fothergill Cooke and Sir Charles Wheatsone had also come up with their own telegraph system. They received a patent, but it was Morse's design and code that stole the day.

As a young man Gray supported himself as a carpenter while attending Oberlin College, where he became enraptured with electricity. He had the bones of an inventor, and began tinkering. In 1867 he got a patent for his design of an improved telegraph relay that adapted to the variable insulation that coated the lines, the first of his more than seventy patents. In 1869 Gray went on to found the telegraph supply company Gray & Barton Co. with his partner Enos M. Barton in Cleveland, Ohio. One of their chief customers was the Western Union Telegraph Company, who then went on to buy Gray & Barton Co. outright and change their name to Western Electric. Gray continued to work and invent for them until 1874, when he went independent and set about work on developing the telephone.

Just as Morse, Fothergill, and Wheatstone were inventing the telegraph, Gray and Alexander Graham Bell both came up with designs for the telephone at the same time, and it wasn't just them who were at work transmitting voices over wire. Both Bell and Gray used liquid transmitters in their experiments with voice transmission over wire. Each inventor developed the work independently, and in fact Gray supposedly arrived at the patent office to file his apparatus "for transmitting vocal sounds telegraphically" just two hours after Bell did. After years in the courts, it was Bell's patent that the lawyers held up in a number of decisions as the first telephone. Though Elisha Gray may only be considered a kind of begrudged step-father in terms of telephony, it is clear that the electric synthesizer is the fruit of his seed.

While working for Western Union Telegraph, Gray had been obsessed with solving the problem of creating a "multiple telegraph," a way to transmit a number of messages over the same wire. He needed to figure out a way to control sound that was coming from an electromagnetic circuit. This led him to the invention of a basic oscillator made of steel rods whose vibrations were created and transmitted over a telegraph line. The instrument consisted of a number of single-tone oscillators that could play over a range of two octaves. Each tone was controlled with a separate telegraph key. This Electro-Harmonic Telegraph was one of the earliest electronic musical instruments, and used keys like on a piano or organ to open and close a circuit creating a similar buzzing tone as used in Morse Code. His keyboard spanned the pitch range of two octaves making the tone higher and lower in places than it would have normally been when just used for sending a telegram. An operator could now potentially dash out simple melodies over the wires, for the telegraph operator on the other end to hear. The use of oscillators prefigured their later extensive use in making music, and the multi-tone transmitters over the telegraph line prefigured the use of multi-tone dialing in later telephony.

After giving several private demonstrations of the instrument, he gave a public performance at the Presbyterian Church of Highland, Illinois, on December 29, 1874. A newspaper announcement stated that it transmitted "familiar melodies through telegraph wire." In July of 1875, Gray was granted a patent for his "Electric Telegraph for Transmitting Musical Tones."

In later models of the instrument he added a simple diaphragm speaker that amplified the tones to a louder volume.

In William Peck Banning's 1946 book, *Commercial Broadcasting Pioneer: The WEAF Experiment 1922-1926*, he wrote that "historians of the future may conclude that if there was any 'father' of broadcasting, perhaps it was the telephone itself." Today those of us with access to cell phones and data plans tend to take things like streaming music, news, on-demand videos, and FaceTime for granted. Yet the impulse to do more than just talk over the wires has been part of the spirit of telephony since its earliest days. In the 1890s, the telephonic playground was still in its infancy and commercial applications for the technology could have gone in many different directions. During this time entrepreneurial types were coming up with creative experiments for using telephones as a news delivery system or for musical entertainment.

Two years after Elisha Gray's playing of the musical telegraph, other folks decided it would be a swell idea to transmit music concerts by way of the commercial telegraph lines, done initially for the entertainment of the operators. In 1881 the first "stereo" concert was given via telephone when French inventor Clément Ader used dual lines to pass music from a local theater to two separate phone receivers. At the time this was dubbed "binauriclar audition," a name that, for some reason, didn't stick. Later in 1890, AT&T was at work on a service to provide music over the phone for subscribers to listen to at mealtimes. Though there were some issues with sound quality, AT&T stated that "when we have overcome this difficulty we shall be prepared to furnish music on tap," with an ultimate ambition of essentially "streaming" music, lectures, and various audio entertainments to all the cities of the East Coast.

While these types of efforts didn't take hold Stateside, a few in Europe did. The first regular streaming service was an outgrowth of Clément Ader's work, known as the Paris Theatrophone. A subscription-based phone service launched in the 1890s, the "Theatraphonic network" provided Parisians with "programs dramatic and lyrical" and held its own until 1932. Meanwhile in Hungary, the concept of a telephone newspaper caught on via the Budapest Telefon Hirmondo, which began service in February of 1893. It included news reports, original fiction, and other entertainment. Still going strong in 1925, it added a radio station while still offering a telephone relay to customers all the way until 1944.

The Telharmonium

It was during this early phase of tele-technology that Thaddeus Cahill obsessed over and created what must be considered the ultimate behemoth of a musical synthesizer, the Telharmonium, a type of electrical organ. Cahill, like Elisha Gray, had also been a student at Oberlin College, where he studied the physics of music. Cahill took inspiration from Gray's earlier musical telegraph, but thought it could be much improved, and intended to create an instrument to be played over the phone lines. With amplifiers not yet invented, the phone receiver was still the only available technology that could make electrical sounds audible. Rather than use an oscillating circuit, the Telharmonium implemented an early form of additive synthesis via mechanical means using tonewheels, an apparatus that generates musical notes, and alternators, a device that turns mechanical movement into electricity.

Among his other legacies, Thaddeus Cahill is credited with coining the phrase "synthesizer" for describing his instrument, which was patented in 1897. Five years later he founded the New England Electric Music Company with two partners, but it wasn't until 1906 when the Telharmonium, or Dynamaphone as it was also called, was first demonstrated. The instrument was a true boat anchor. The Mark I version weighed in at a hefty seven tons and could be considered light compared to the Mark II and III, which weighed around 200 tons and took up thirty train box cars when shipped to New York for assembly in what Cahill called his Music Plant. Featuring a sixty foot chassis and 145 rotors, with each rotor producing a pitch, the instrument looked like a power generator and took up an entire floor on 39th Street and Broadway in New York City. The machine itself put out 670 kilowatts of power and required massive energy inputs. The sound of the electrical hum and machinery in the Music Plant alone might be enough to send noise music fans into a state of rapturous ecstasy.

One floor up from the instrument was Telharmonic Hall, a concert space where the Telharmonium was controlled and played. Two to four musicians could sit at the controls to play inside the listening hall. It was a unique arrangement of four keyboard banks each with eighty-four keys. Before the minimalist composers La Monte Young and Terry Riley brought it back into the fold of Western music in the 1960s, it was possible to play the Telharmonium using the tuning style of "just intonation." Just intonation differs from equal temperament—and was developed for keyboard instruments so that they could be played in any scale or key—in that it occurs naturally as a series of overtones where all the notes in a scale are related by rational numbers. Through additive synthesis, and the ability to control timbre, harmonics, and volume, the Telharmonium was extremely flexible in its sound, if cumbersome in its sprawling apparatus.

Though there was no channel separation, the Telharmonic Hall was fitted with eight telephone receivers augmented with paper horns to amplify the sound, which were arrayed behind ferns, columns, and furniture. An electrician at the company suggested splicing the current from the Telharmonium into the electric arc lamps hanging from the ceiling, which then resonated at the same audible frequency as was being played to create a "singing arc." The Telharmonium could also be piped to any phone number connected to the AT&T system.

Engineer Thomas Commerford Martin described the new sounds of the Telharmonium as an alliance of electricity with music, writing that Cahill had "devised a mechanism which throws on the circuits, manipulated by the performer at the central keyboard, the electrical current waves that, received by the telephone diaphragm at any one of ten thousand subscribers' stations, produce musical sounds of unprecedented clearness, sweetness, and purity."

Thaddeus Cahill had ambitious plans for his Telharmony -the kind of music played on a Telharmonium. He advocated that a form of "electric sleep-music" could be tapped at any time for the cure of modern nervous disorders. The electric drones could also be used to relieve boredom in the workplace. Both of these uses are how many people use ambient music today. But his plans were not to bear fruit in the manner he thought. His massive instrument sometimes caused interference or crosstalk on the phone lines, with electronic music interrupting business and domestic conversations. It also required vast amounts of power. When vacuum tubes started to appear and in the 1920s, less expensive electronic instruments started being built. Then, with the advent of broadcast radio, many of these types of ventures ceased to be profitable. It is unfortunate that no known recordings of the Telharmonium exist.

In the 1930s, American engineer Laurens Hammond patented the electrical amplified organ. Hammond organs generated sound by creating an electric current from a metal tonewheel that rotated near an electromagnetic pickup. It was essentially a smaller and more economical version of Cahill's Telharmonium, whose patent had not yet run out. It seems no coincidence then that when Cahill passed away in 1934, the Hammond organ went into production the following year.

Thaddeus Cahill's work and his seminal place in the history of electronic music has since been recognized, even if the instrument itself had long ago been recycled for scrap metal.

2: A Wireless Fantasy

The Sounds of the Arc

"A light bulb creates an environment by its mere presence," wrote influential philosopher Marshall McLuhan in 1964, using the metaphor of a light bulb to frame his entire understanding of media theory. It is only appropriate then that a key development in radio was made by inventors exploring the musical potential of early electrical lighting.

The first source of electrical lighting was the arc lamp. Invented by Humphry Davy in the first decade of the nineteenth century, the arc lamp created light from the electricity passing between two carbon electrodes in free air. To ignite a carbon lamp the rods were touched together, allowing a low voltage to strike the arc, then drawn apart to allow the electric current to flow between the gap. This first means of electrical lighting also became the first commercial use for electricity beginning around 1850, though it didn't really take off until the 1870s when regular supplies of power became available.

During the 1880s major advances in the technology occurred that helped spread the adoption of the arc lamp and a number of companies became involved in there manufacture as they began to be used for lighting streets and other public places. The only problem was a feature of the light source many folks found disagreeable: the audible power-frequency harmonics caused by the arcs' negative resistance. In other words, noise. Famed inventor Nikola Tesla was one of the men who set out to solve this problem and in 1891, he received a patent for an alternator running at 10,000 cycles per second that would suppress the undesirable sounds of humming, hissing, and howling emitted by the arc lamps.

Yet there is a musicality to noise, later to be explored in depth by the likes of the Futurists, John Cage, and Throbbing Gristle, among others. As collective electrical activity tarnished the relative silence with audio pollution, it also opened the door to new ways of perceiving these sounds—not just as a noisome nuisance, but as vibrations with their own unique essence.

Tesla's invention must have been impractical or just never caught on, because in 1899 London-based electrical engineer William Duddell had been appointed to tackle the problem of the lamps' dissonant electrical noise. Duddell was an illuminated man, no pun intended, and he took a different angle than Tesla. Instead of suppressing the sounds, he transformed them into music. In the course of his experimentation, Duddell found that, by varying the voltage supplied to the lamps, he could control the audible frequencies by connecting a tuned circuit that consisted of an inductor and a capacitor, both components that store energy, across the arc. The arc, in a state of negative resistance, became excited by the audio frequency oscillations from the tuned circuit at its "resonant frequency," the natural frequency where the circuit would vibrate at its highest amplitude. This could be heard as a musical tone.

Duddell used another one of his inventions, the oscillograph, to analyze the particular conditions necessary for producing the audible oscillations of different electronic frequencies. He demonstrated his invention before the London Institution of Electrical Engineers by wiring up a keyboard to make different tones from the arc, and being a patriotic fellow, even played a rendition of "God Save the Queen." His device came to be known as the "Singing Arc" and was one of the first electronic oscillators. It was noted that arc lamps on the same circuit in other buildings could also be made to sing, leading the engineers to speculate that music could be delivered over the lighting network, but this never became a reality, with the exception of Thaddeus Cahill's Telharmonium. Duddell toured his instrument around Britain for a time but his invention was never capitalized on and remained a novelty.

William Duddell's Singing Arc had been very close to becoming a radio. The spark-gap transmitter invented by Italy's Guglielmo Marconi had already been demonstrated in 1895, yet Duddell thought it was impossible to leverage his Singing Arc to produce radio frequencies instead of audio frequencies. The alternating current (AC) in the condenser was smaller than the supplied direct current (DC), so the arc never extinguished during an output cycle, making it impractical to use as a radio frequency (RF) transmitter. With this setup, it was not possible to reach the high frequencies required for transmission of radiotelegraphy. If he had managed to increase the frequency range and attached an antenna, his invention could have become a continuous wave (CW), or Morse Code transmitter.

Duddell's oscillator was left for other experimenters to improve upon. This was done by Danish physicists Valdemar Poulsen and P.O. Pederson. In 1903 they patented the Poulsen arc wireless transmitter, the first device to generate continuous waves and one of the first pieces of technology to transmit through amplitude modulation.

Without William Duddell attempting to make the sounds of arc lamps more pleasant, Valdemar Poulsen may have never thought of adapting the original invention for use in the radio frequency spectrum. His design helped radio transmissions become a more reliable form of contact between two points, and they were installed on ships at sea to stay in contact with stations on shore. This type of transmitter was used for radio work around the world until the 1920s when vacuum tube technology came along to reinvent the way radio was done.

The Audio Piano and Wireless Fantasies

No man works in a vacuum. Before the industry of radio got off the ground, it had been customary for researchers to use each other's discoveries with complete abandon. As technical progress in the field of wireless communication moved from the domain of scientific exploration to commercial development, financial assets and intellectual property came to be at stake, and rival inventors soon got involved in one of the great American pastimes: lawsuits. The self-styled "Father of Radio" Lee de Forest was involved in a number of infringement controversies, the most famous of which involved his invention of the "audion" (from *aud*io and *ion*ize), an electronic amplifying vacuum tube.

It was Thomas Edison who first produced the ancestor of what became the audion. While working on the electric light bulb, Edison noticed that one side of the carbon filament behaved in a way that caused the blackening of the glass. Working on this problem, he inserted a small electrode and was able to demonstrate that it would only operate when connected to the positive side of a battery. Edison had formed a one-way vacuum tube, or valve, as it is also called. This electrical phenomenon made quite the impression on another experimenter, English physicist John Ambrose Fleming, who brought the device back to life twenty years later when he realized it could be used as a radio wave detector.

At the time Fleming was working as an advisor for Guglielmo Marconi. It occurred to him that if he connected an Edison effect bulb to an antenna, grounded the filament, and placed a telephone within this circuit, that he could reduce the frequencies so that the receiver would put out audio from the effect of the waves. Fleming made these adjustments and substituted a metal cylinder for Edison's flat plate, finding that the sensitivity of the device was improved by increasing electronic emissions. This great idea in wireless communication was called the Fleming valve.

Fleming had patented this two-electrode tube in England in 1904, before giving the rights to the Marconi Company, which took out American patents in 1905. Meanwhile, American inventor Lee de Forest had read a report from a meeting of the Royal Society where Fleming had lectured on the operation of his detector. De Forest immediately began experimentation with the apparatus on his own and found himself dissatisfied. Between the cathode and anode, or positive and negative sides, he added a third element made of a platinum grid that received current coming in from the antenna. This addition would soon transform the field of radio, setting powerful forces of electricity—as well as litigation—into motion.

The audion increased amplification on the receiving side, but radio enthusiasts were doubtful about the ability of the so-called "triode" tube to be used with success as a transmitter. Finding himself in financial trouble after various scandals in the wireless world, de Forest was persuaded to sell his audion patent in 1913.

De Forest found a nemesis in fellow American inventor Edwin Howard Armstrong, who had been fascinated by radio since his boyhood. When de Forest was thirty-two years old, Armstrong was just fifteen and already an amateur radio operator. Experimenting with the early de Forest audions, which were not perfect vacuums (de Forest had mistakenly thought a little bit of gas left inside was beneficial to receiving), Armstrong took a close interest in how the audion worked and developed a keen scientific understanding of its principles and operation. By the time he was a student at Columbia University in 1914, he was doing intensive work using an oscillograph to make comprehensive studies based on his fresh and original ideas. Following the muse of invention, he discovered regenerative feedback, another coup for the growing wireless industry, revealing that when feedback was increased beyond a certain point, a vacuum tube would go into oscillation and could be used as a continuous wave transmitter. Thus, Edwin Armstrong received a patent for the regenerative circuit.

In turn, Lee de Forest claimed he had already come up with the regenerative principle in his own lab, and so the lawsuits began, and continued for twenty years, with victories that alternated as fast as electric current. In 1934, the Supreme Court made a final decision in the favor of de Forest, saying he had the right in the matter. Armstrong, however, would achieve lasting fame for his superheterodyne receiver, invented in 1918.

Around 1915, de Forest had used heterodyning—where two signals are merged to create a third—to create a musical instrument out of his triode valve. His Audion Piano was to be the first instrument created with vacuum tubes (i.e. valves), which became the basis for nearly all future electronic instruments until the invention of the transistor in 1947.

Looking like a shrunken version of today's electronic pianos, the Audion Piano consisted of a single keyboard manual and used one triode valve per octave. The set of keys allowed one monophonic note to be played per octave. Out of this limited palette, the instrument created variety by processing the audio signal through a series of resistors and capacitors that vary the timbre. The Audion Piano was notable for its spatial effects, prefiguring the role electronics would play in the spatial movement of sound. The output could be sent to a number of speakers placed around the room to create an enveloping ambiance. De Forest later planned to build an improved version with separate tubes for each key to give it full polyphony, but it is not known if it was ever created.

In his grandiose autobiography, Lee de Forest described his instrument:

> *...sounds resembling a violin, cello, woodwind, muted brass and other sounds resembling nothing ever heard from an orchestra or by the human ear up to that time – of the sort now often heard in nerve racking maniacal cacophonies of a lunatic swing band. Such tones led me to dub my new instrument the 'Squawk-a-phone'.... The Pitch of the notes is very easily regulated by changing the capacity or the inductance in the circuits, which can be easily effected by a sliding contact or simply by turning the knob of a condenser. In fact, the pitch of the notes can be changed by merely putting the finger on certain parts of the circuit. In this way very weird and beautiful effects can easily be obtained.*

In 1915, when an Audion Piano concert was held for the National Electric Light Association, a reporter wrote the following:

> *Not only does De Forest detect with the Audion musical sounds silently sent by wireless from great distances, but he creates the music of a flute, a violin or the singing of a bird by pressing a button. The tone quality and the intensity are regulated by the resistors and by induction coils... You have doubtless heard the peculiar, plaintive notes of the Hawaiian ukulele, produced by the players*

> *sliding their fingers along the strings after they have been put in vibration. Now, this same effect, which can be weirdly pleasing when skillfully made, can be obtained with the musical Audion.*

Lee de Forest's accomplishments are many. He had been an avid experimenter in the early days of broadcasting with his 2XG station and as an inventor he held over three hundred patents. He put a great deal of energy into creating an optical sound-on-film system called Phonofilm, yet of all his technical achievements, he called the Audion amplifying vacuum tube his "greatest invention," and though he was arrogant and contentious, it's hard to argue. Vacuum tubes changed the entire field of electronics and telecommunications for some forty years enabling their growth and further development. Long-distance telephone calls, radio broadcasting and television all benefited from the warm glow emitted by his tubes. Even the first electronic computers used vacuum tubes to do their mighty calculations, and many of the coming generation of electronic instruments got their rich sound from the tubes that gave them a spark of life, including the Theremin.

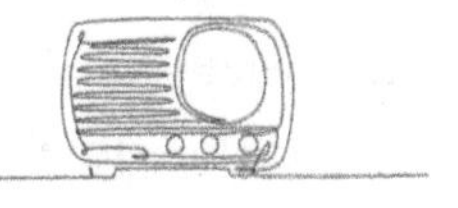

3: VIBRATIONS FROM THE AETHER

Lev Theremin and the Vibrations of the Aether

The sound of the theremin has become synonymous with the spectral and spooky sci-fi horror flicks of the 1940s and '50s. Its trilling oscillations conjure up images of flying saucers made from hubcaps and fishing line. When most folks hear and see the theremin, they tend to think of it as little more than a novelty or scientific amusement. While it may have fallen out of favor in horror movie soundtracks, it has remained a mainstay within the field of electronic music, still distinguished among other musical instruments by being played without actually touching the instrument itself. In this way it is similar to an aeolian harp—a box-like string instrument played by the blowing of wind—and both seem to conjure forth sounds from other planes. To the radio and electronics buff, the theremin is worth exploring as a way of learning about electromagnetic fields, and it is a great example of the creative use of the heterodyning effect, where two signals are mixed together to make a third, for artistic purposes.

The inventor of the theremin, or "etherphone" as it was first dubbed, was Leon Theremin. Born in Russia in 1896, a few years before Marconi achieved wireless telegraphy, he spent his childhood reading the family encyclopedia and was fascinated by physics and electricity. At five he had started playing piano, and by nine had taken up the cello, an instrument that has an important influence on the way theremins are played. After showing promise in class, he was asked to do independent research with electricity at the school physics lab, where he began an earnest study of high-frequency currents and magnetic fields, alongside optics and astronomy. It was around this time that Theremin met Abram Ioffe, a rising physicist whom he would work under in a variety of capacities, but his studies in atomic theory and music were overshadowed by the outbreak of the first World War. In 1916 Theremin was summoned by the draft and moved to St. Petersburg, which had just been renamed Petrograd at the onset of the war, where his electrical experience saved him from the front lines. He was placed in a military engineering school, where he landed in the Radio Technical Department to do work on transmitters and oversee the construction of a powerful and strategic radio station, which later had to be disassembled, with Theremin overseeing the blowing up of a 120-meter antenna mast. Amidst all this, he also worked as a teacher instructing other students to become radio specialists.

As Theremin's reputation grew among Russian engineers and academic scientists, he was eventually asked to go and work with his mentor Abram Ioffe at the Physico-Technical Institute, where he became the supervisor of a high-frequency oscillations laboratory, and his first assignment was to study the crystal structure of various objects using X-rays. At the time he was also experimenting with hypnosis, leading Ioffe to suggest he take his findings on trance-induced subjects to psychologist Ivan Pavlov. Though Theremin resented radio work in preference for his love of exploration of atomic structures, Ioffe pushed him to work more systematically with radio technology. So, in the early 1920s, Theremin busied himself thinking of novel uses for the audion tube.

His first project involved the exploration of the human body's natural electrical capacitance to set up a simple burglar alarm circuit that he called the "radio watchman." The device was made by using an audion as a transmitter at a specific high frequency directed to an antenna. This antenna only radiated a small field of about sixteen feet. The circuits were calibrated so that when a person walked into the radiation pattern it would change the capacitance, causing a contact switch to close and setting off an audible signal.

Theremin was next asked to create a high-frequency oscillator for measuring the dielectric constant of gasses in a variety of conditions. For this he made a circuit and placed a gas between two plates of a capacitor, with changes in temperature measured by a needle on a meter. This device was so sensitive it could be set off by the slightest movement of the hand, and was refined by adding an audion oscillator and tuned circuit. The harmonics generated by the oscillator were filtered out to leave a single frequency that could be listened to on headphones.

As Theremin played with this tool, he noticed again how the presence of his movements near the circuitry were registered as variations in the density of the gas, and now measured by a change in the pitch. Closer to the capacitor the pitch became higher, while further away it became lower. Shaking his hand created vibrato. His musical self, long dormant under the influence of war and communism, came alive, and he started to use this instrument to tease out the fragments he loved from his classical repertoire. Word quickly traveled around the institute that Theremin was playing music on a voltmeter, and Ioffe encouraged him to refine what he had discovered—the capacitance of the body interacting with a circuit to change its frequency—into an instrument.

To increase the instrument's range and have greater control of the pitch, he employed the heterodyning principle. He used two high-frequency oscillators to generate the same note in the range of 300 kilohertz, which lies beyond the range of human hearing. One frequency was fixed, the other was variable and could move out of sync with the first. He attached the variable circuit to a vertical antenna on the right-hand side of the instrument. This served as one plate of the capacitor while the human hand formed another. The capacitance rose or fell depending on where the hand was in relation to the antenna. The two frequencies were then mixed into a beat frequency, or an interference pattern between sounds where there is a slight difference of frequencies. To play a song, the hand is moved at various distances from the antenna creating a series of beat frequency notes.

To refine his etherphone further, he designed a horizontal loop antenna that came out of the box at a right angle. Connected to carefully adjusted amplifier tubes and circuits, this antenna was used by the other hand to control volume. This newborn instrument had a range of four octaves and was played in a similar manner to the cello, as far as the motions of the two hands were concerned. After playing the instrument for his mentor, he performed a concert in November of 1920 to an audience of spellbound physics students. In 1921 he filed for a Russian patent on the device.

Leon Theremin's skill at invention was not lost on the Soviet machine. Not long after his musical instrument was patented, the radio watchman security device it was in part based on started being employed to guard the treasures of gold and silver Lenin had plundered from church and clergy. The watchman was also on guard at the state bank, with the installation of these early electronic traps taking him away from his primary interest in scientific research. Just as he was approaching the limits of his frustration, his mentor at the Institute gave him a new problem to solve: that of "distance vision," or the transmission and reception of moving images over the airwaves. The embryonic idea for television was in the air at the time, but no one had figured out how to make it a reality. The race was on and the Soviets wanted to be first to crack the puzzle.

Having researched the issue extensively in the published literature, Theremin was ready to apply the powers of his mind towards a solution. Being that parts weren't always readily available in the Soviet Union, some were smuggled in and others had to be scavenged from flea markets, the latter a process very familiar to today's radio and electronics junkies. By 1925 he had created a prototype from his junk box using a rotating disk with mirrors that directed light onto a photocell. The received image had a resolution of sixteen lines, which was enough to make out the shape of an object or person but not the identifiable details. Fine tuning the instrument over the next year, he doubled the resolution to thirty-two lines and then, using interlaced scanning, to sixty-four. Having created a rudimentary "Mechanism of Electric Distance Vision," he demonstrated the device and defended his thesis before students and faculty from the physics department at the Polytechnic Institute. Theremin had built the first functional television in Russia.

After the first World War, Theremin embarked to Europe and then America, where he lived for just over a decade, engaging the public, generating interest in his musical instrument, and working with RCA. As Hitler gathered power back in Europe, he was anxious about the encroaching war and returned home to the Soviet Union in 1938. He barely had time to settle back down when he was sent to the Kolyma gold mines for forced labor for the better part of a year. This was done as a way of breaking him, a fear tactic that could be held over his head if he didn't cooperate, before the state found better uses for him. He was picked up by the police overlord Lavrenti Beria, who sent him to work in a secret laboratory that was part of the Gulag camp system. One of his first jobs was to build a radio beacon whose signals would help track down missing submarines, aircraft, and smuggled cargo.

As World War II wound to a close, the Cold War began to dawn, and the USSR was on the offensive, trying to extend its reach and gather intelligence on such lighthearted subjects as the building of atomic bombs. In their efforts at organized espionage, the Soviets sifted for all the data they could get from foreign consulates. Having succeeded with his beacon, Theremin was given another assignment. This time the goal wasn't to track down cargo or vehicles but to intercept US secrets from inside Spaso House, the residence of the US Ambassador in Moscow. Failure to do the bidding of his boss would mean a return to the mines, and his boss had high demands for the specifications of the bug he was to plant. The proposed system could have no microphones and no wires, and was to be encased in something that didn't draw attention to itself.

The bug ended up being put inside a wooden carving of the Great Seal of the United States and was delivered by a delegation of Soviet Pioneers, a group similar to the Boy Scouts, on July 4, 1945. Deep inside this "gesture of friendship" was a miniature metal cylinder with a nine-inch antenna tail. The device was passive and therefore undetectable by the X-rays used at Spaso House in their routine scans. It only activated when a microwave beam of 330 megahertz was directed at the seal from a nearby building. There was a metal plate inside the cylinder that, when hit with the beam, resonated as a tuned circuit. Below the beak of the eagle, the wood was thin enough to act as a diaphragm and the vibrations from it caused fluctuations in the capacitance between the plate and the diaphragm, creating a microphone. The modulations this produced were then picked up by the antenna and transmitted out to the receiver at a Soviet listening post. Using this judiciously, the Soviets were able to gain intelligence to aid them in a number of strategic decisions. Today the Great Seal bug is considered to be a grandfather to radio frequency identification (RFID) technology.

This wasn't the last time Theremin was asked to develop an apparatus for wireless eavesdropping. For the next job, an operation code named Snowstorm, his overseers upped the ante. This time, no device could be planted in the site targeted for surveillance, leading Theremin to find inspiration in optics. Knowing that window panes in a room vibrate slightly when people talked, he needed a method to detect and read those vibrations from a distance. Resonating glass contains many simultaneous harmonics and it would be difficult to find the place of least distortion to get a voice signal from. Then there was the obstacle of reinterpreting the signal back into a speech pattern. Using an infrared beam focused on the optimum spot and catching its reflection back in an interferometer (an optical tool that uses light interference patterns) tricked out with a photo element, he was able to pick up communications. Back at his monitoring post he used his equipment and skills to reduce the large amounts of noise from the signal.

A few years later Leon Theremin was released from his duties at the lab, but was kept on a tight leash and not allowed to leave Moscow, even as his instrument continued to create waves around the world.

The Chance Meeting of a Torpedo and a Player Piano Roll

In the conventional transmission and receiving of signals, the frequency does not change over time, except for small fluctuations due to certain types of modulation that can cause a signal to slightly drift or expand. For obvious, practical reasons the signal is kept on a single frequency so two people communicating can exchange information, or so a listener in the broadcast bands knows exactly where to go to find his favorite station. That is all fine and dandy for typical uses of radio, but as the medium developed, the inventors and researchers who expanded the state of the art found a couple of hitches that made it problematic for certain types of signals to remain parked on one frequency. The first was interference caused by deliberate jamming on the desired frequency, as has been done by various governments trying to block foreign broadcasts of propaganda. This category also includes non-malicious interference coming from transmissions on nearby frequencies. The second issue with using only one frequency in a communication is when the information being transmitted is of a sensitive nature.

Constant-frequency signals are easy to intercept. The military and others can make use of codes and encryption to veil transmissions on single frequencies, but codes can be broken. Radio researchers found that another layer of communication security could be added by the use of a collection of techniques known as spread spectrum radio, a method where the frequency of the signal is intentionally varied. Frequency-hopping was the first of these techniques to be established.

Though attributed to multiple inventors, the first patent for frequency hopping was granted to, of all people, Hollywood actress

Hedy Lamarr and composer George Antheil in 1942 for their "Secret Communications System," designed to protect Allied radio-guided torpedoes from being jammed by the Axis powers. Both Lamarr and Antheil are most remembered for their main fields of activity, movies and music, but they each had a touch of the polymath inside of them, and their other passions allowed them to make a significant advancement in the art and science of transmission.

Born in Vienna in 1914, Hedy Lamarr was just eighteen when she married the first of her six husbands, Friedrich "Fritz" Mandl, a wealthy munitions manufacturer whose weapon systems later gave her inspiration for the patent. During this time she had started her film career in Czechoslovakia with the 1933 movie, *Ecstasy*, which became controversial for its frank depictions of nudity and sexuality. Hubby Mandl got a bit ticked off by these movie scenes and attempted to stop Lamarr from continuing her career as an actress.

Both Mandl and Lamarr had Jewish parents, but Mandl also had business ties with the Nazi government, to whom he sold his weapons. Mussolini and Hitler were among those who attended the lavish parties Mandl hosted at their Schloss Schwarzenau castle, and Lamarr would accompany him to his meetings where she got to associate with scientists and professionals involved in military technology. It was at these conferences where her interests in inventing and applied science were first sparked.

As her marriage grew unbearable, Lamarr decided to flee to Paris, where she met movie mogul Louis B. Mayer, who was scouting for talent. With all the trouble brewing in Europe, Mayer found it easy to persuade Lamarr to move to Hollywood, where she arrived in 1938 and began work on the film *Algiers*.

It was during the height of the war and her career that she herself grew bored with acting. Cast for her beauty, rather than her talent and ability, Lamarr had complained that the roles given to her required little challenge in terms of technique or the delivery of lines and monologues. Stifled by the lack of more demanding roles, she found an outlet for her intellectual capacities through the hobby of tinkering and inventing, which was nurtured by her friendship with aviation tycoon and germaphobe Howard Hughes.

Before finding success in Hollywood scoring films, avant-garde composer George Antheil was part of the Lost Generation and, like contemporaries such as Ernest Hemingway, had moved to Europe after the horrors of the first World War to live a bohemian life amidst the cafes and salons of 1920s Paris. It was during this period when he composed his best-known work, "Ballet Mecanique", which began its life as an accompaniment to the Dadaist film of the same name made by Fernand Léger and Dudley Murphy, with cinematography by Man Ray. The techniques Antheil developed in this composition were to be key to the success of the frequency hopping patent he would share with Lamarr.

"Ballet Mecanique" was scored to use a number of self-playing "player pianos." He described their effect as "All percussive. Like machines. All efficiency. No LOVE. Written without sympathy. Written cold as an army operates. Revolutionary as nothing has been revolutionary." There are no human dancers to this ballet, just mechanical instruments. Antheil's original conception was to use sixteen specially synchronized player pianos, two grand pianos, electronic bells, xylophones, bass drums, a siren, and three airplane propellers. There were a number of difficulties involved in this setup that broke away from traditional orchestral arrangements. The synchronization of the player pianos proved to be the largest obstacle. Consisting of periods of music and interludes of relative silence created by the droning roar of airplane propellers, Antheil described it as "the rhythm of machinery, presented as beautifully as an artist knows how."

Besides composing, Antheil was a writer and fierce patriot. He was a member of the Hollywood Anti-Nazi League and wrote a book of predictions about the conflict titled *The Shape of the War to Come*. He also penned a newspaper column on relationship advice that was nationally syndicated and he fancied himself an expert on the subject of female endocrinology. His interests in this area was what first brought him into contact with Lamarr, who had sought him out for advice on how she might, shall we say, enhance her upper torso. After he proposed that she could make use of glandular extracts, their conversation (somehow) turned to the kind of torpedoes being used in the war.

By then Lamarr was herself a staunch supporter of her adopted country, though she didn't become a naturalized US citizen until 1953. Using knowledge she gained from her first marriage with the munitions manufacturer, she had the insight that radio controlled torpedoes would excel in the fight against the Axis powers. The radio signals could easily be jammed though, and the torpedo sent off course. Working with Antheil, she devised their "Secret Communications System."

The action of composing for the player pianos helped Antheil with one of the aspects of creating their system, which had a striking resemblance to the still top secret SIGSALY system, described in the next chapter. It is best described in the overview of their patent number 2,292,387:

> *Briefly, our system as adapted for radio control of a remote craft, employs a pair of synchronous records, one at the transmitting station and one at the receiving station, which change the tuning of the transmitting and receiving apparatus from time to time, so that without knowledge of the records an enemy would be unable to determine at what frequency a controlling impulse would be sent.*

They had the idea they could use the same kind of records that were used for in player pianos that consisted of long paper rolls that had holes punched in them in a number rows along the record. Player pianos usually had 88 rows of these perforation. In their system these rows would not be playing notes, but would instead allow the use of 88 different radio carrier frequencies. The transmitted signal, guided by this perforated record, would hop from one frequency to another. These kinds of records could be made very long, and could be played at different speeds in order to run for enough time that a remote control could guide a torpedo to its target. There would need to be a set of two, one at each station for transmission and reception and the two stations would be synchronized by the same kind of motors used to calibrate clocks and chronometers. Positions would be able to be corrected by sending out a series of synchronous impulses for realigning the machinery, as similar systems were already at use in telegraphy and television.

Although the US Navy did not adopt their technology until the 1960s, the principles of their work continue to live on and are now used in everyday devices such as Wi-Fi and Bluetooth technology. Spread spectrum systems are also used in the unregulated 2.4 gigahertz band and on some walkie-talkies that operate in the 900 megahertz portions of the radio spectrum. Other spread spectrum techniques include direct-sequence spread spectrum, time-hopping spread spectrum, and chirp spread spectrum.

In 2008, Elyse Singer wrote the script for an off-Broadway play, *Frequency Hopping*, that spotlights the lives and scientific work of Hedy Lamarr and George Antheil, winning a prize for best new play about science and technology. Lamarr and Antheil's pioneering work eventually led to their posthumous induction into the National Inventors Hall of Fame in 2014.

PART II

THE VOICE OF THE BELL

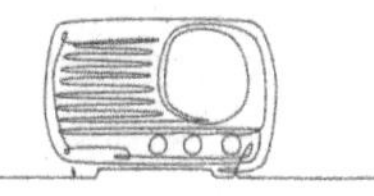

4: ENCIPHERED SOUNDS

Auditory Perception and Articulation

It could be said that speech recognition and synthesis systems began in the nineteenth century, when sound waves were first converted into electrical signals. From that point, a number of researchers at Bell Labs became interested in various aspects of speech.

Harvey Fletcher was one of these researchers. Though he hadn't intended on becoming a physicist and inventor when he was starting out, his knack for physics and mathematics led him to leave his hometown of Provo, Utah, for the University of Chicago. There he was taken under the wing of physicist Robert A. Millikan, whose groundbreaking work on how to measure the charge of an electron, fundamental to the growth of electronics and broadcasting technologies, eventually earned him a Nobel Prize.

After school, Fletcher hitched his wagon to the Western Electric Company in New York. To gain perspective, he got his hands dirty climbing telephone poles, operating switchboards, and installing phones. Then he started formulating his research plan. Realizing that the fundamental aspect of the Bell System was the transmission of speech, he felt it was necessary to understand all the different components of the physical body that related to speech and hearing in order to provide the company with a solid body of work on which they could perfect their system. Part of this research included connecting bellows to a vibrating reed that he put in his mouth to make different speech sounds. He would repeat each sound thoroughly until he was sure he understood the exact placement of tongue and shape of lips. This new vein of research was summarized in his first paper on the subject, "On the Relative Difficulty of Interpreting the English Speech Sounds," published in 1920.

From his work with the vibrating reed mechanism, Harvey Fletcher invented an artificial larynx to help those who had lost theirs due to cancer surgeries and other procedures. These were manufactured and sold by the Bell System to help those who had lost the ability to speak to regain their voice.

At Bell Labs, under the auspices of pure research, Fletcher's gifts blossomed. He published fifty-one papers, wrote two books, and had nineteen patents. In particular his two books, *Speech and Hearing* and *Speech and Hearing in Communication*, set the precedent for further work on the clarity of audio.

Harvey Fletcher's work with acoustics enabled the synchronization of sound and image in the still young film industry. Thanks to his research, talking pictures emerged from the silence, and for many years were made under Bell patents, even as Fletcher resisted offers to come to Hollywood.

In the early talkies, there was no surround sound, panning, or spatial movement of audio. Fletcher knew a more immersive experience could be achieved, having worked with Alfred DuPont in creating a binaural system for the board meetings of his company. In 1931, Fletcher worked with colleague Arthur C. Keller and conductor Leopold Stokowski to record and transmit monaural and binaural—or stereo—sound for the Academy of Music in Philadelphia.

In 1933, alongside his colleague Wilden Munson, Fletcher conducted the first research on the frequency response of the human ear. By playing a series of tones, they were able to determine how listeners perceived loudness at different frequencies. From their results they figured out that a listener's perception of pitch varies based on the loudness of the sound. They used the data from these experiments to create the Fletcher-Munson curve, which shows that the most sensitive frequency range in the human ear is between two and five kilohertz, published in their paper, "Loudness, its definition, measurement and calculation," in the *Journal of the Acoustical Society of America*. AT&T used this research to equalize the phone lines and keep the maximum articulation of speech at the sweet spot between two and three kilohertz. Many decades later, influential sound and radio engineer Bob Heil, would use the Fletcher-Munson curve as the basis for his line of ham radio equalizers. In the music world, Heil is more well known for his work with the Grateful Dead, developing their live sound system after their longtime sound man Owsley Stanley was jailed for LSD possession, as well as producing a line of microphones favored by Joe Walsh of The Eagles.

In 1940 Fletcher again worked with Stokowski to make a presentation of stereophonic sound at Carnegie Hall. They recorded stereo music from a three-channel system, with the sounds recorded onto film in a frequency range of 30 hertz to 15,000 kilohertz, with volumes reaching an incredible intensity of 120 decibels, the equivalent of a loud police siren. They used a fourth track to control this massive sonority. The event led *The New York Times* to report, "The loudest sounds ever created crashed and echoed through venerable Carnegie Hall last night as a specially invited audience listened, spellbound, and at times not a little terrified." Although stereo sound had just been unleashed on the public, it wasn't until the late 1950s that stereo technology was adopted in the record industry, and not until the late 1960s that it became standard on most LPs.

Fletcher eventually went on to become the Director of Physical Research at Bell Laboratories, and under his directorship William Shockley, John Bardeen, and Walter Brattain developed the transistor and received the Nobel Prize, which subsequently paved the way for the creation of the semiconductor, which allowed electronics to start becoming miniaturized, paving the way for the microprocessors that are the foundation of today's computers and electronic systems. He also supervised the work that birthed the color TV, advanced medical technology, and ground control communications systems for satellites and spacecraft.

As a whole, Harvey Fletcher's pioneering work on creating crisp audio, clarity of transmission, and the reception of the human voice laid significant groundwork for the field of speech synthesis and recognition.

The Voder and the Vocoder

Homer Dudley, the inventor of the vocoder, was an electronic and acoustic engineer whose work revolved around the idea that human speech is fundamentally a form of radio communication. In his white paper, "The Carrier Nature of Speech," he wrote that "speech is like a radio wave in that information is transmitted over a suitably chosen carrier," a carrier being a waveform encoded with information, in this case spoken language. This realization came to Dudley in October of 1928, when he was otherwise out of commission in a Manhattan hospital bed. Discoveries are often made from playfully messing around with things, either in horseplay or boredom, and Dudley kept himself entertained, just as a kid might, by making weird sounds with his voice through changing the shape of his mouth. He had the insight that his vocal cords were acting as a transmitter of a periodic waveform, a kind of wave that repeats at a constant, with his nose and throat acting as resonating filters and his mouth and tongue producing harmonic content, or formants, these being the distinctive pitch of vowel sounds. He also observed that the frequencies of his voice vibrated at a faster rate than the mouth itself moved.

Once recovered and back at work, these insights went on to have implications for the work he pursued at Bell Laboratories, a true idea factory where money and resources were thrown at any old project that might bear the AT&T monopoly some further advantage in their already sprawling playground of wires and exchanges. Homer thought his discovery might have an application in the area of sound compression, and he made it his ambition to free up some of the phone company's precious bandwidth by packing in more conversations onto the same copper lines.

In radio, a "band" is a given set of radio frequencies, and bandwidth is the frequency range occupied by a modulated carrier signal. In the telephone system, bandwidth refers to the range of frequencies within which the system can produce clarity of speech from one end of a call to another. Bandwidth measured in hertz determines the capacity for the communication channel, in this case the phone line. Bandwidth was important to Ma Bell because if they could get more conversations to fit onto a single phone line, they could make more money. The way to do that was through compression of the voice signals. This became Dudley's special mission.

The human voice ranges between 100 hertz and 17 kilohertz. Traditional telephone calls limited the audio frequencies to between 300 hertz to 3.4 kilohertz. Figuring out a way to further compress the bandwidth of speech onto the bandwidth of the phone line without losing intelligibility or information, was a challenging project. Homer Dudley was given a corner and allowed to work, and he devoted himself to this obsession.

He exploited his research in the invention of the vocoder, or VOice CODER, first demonstrated at Harvard in 1936. It works by measuring how the spectral characteristics of speech—pitch, amplitude, noise, and other aspects of the sound—change over time. The signal going into the microphone is divided by filters into a number of frequency bands, which allows the vocoder to reduce the information needed to store speech to a series of numbers representing each band. On the output end, at the speaker or headphone, the vocoder reverses the process to synthesize speech electronically. Information about the instantaneous frequency of the original voice signal is discarded by the filters, giving the end result robotic and dehumanized characteristics.

There is usually an unvoiced band or sibilance channel on a vocoder for frequencies outside the analysis bands for typical talking but still important in speech. These are words starting with the letters s, f, ch, or other sibilant sounds. These get mixed with the carrier output for increased clarity, resulting in more recognizable speech, but with the characteristic roboticized vocoder sound. Some vocoders have a second system for generating unvoiced sounds, using a noise generator instead of the fundamental frequency.

To better demonstrate the speech synthesis ability of the decoder part of his invention, Dudley created another instrument, the voder, or Voice Operating Demonstrator. This was unveiled during the 1939 World Fair in New York, where author Ray Bradbury was among the attendees who witnessed it firsthand. Another prominent spectator was Vannevar Bush, one of the mentors of "the father of information theory" Claude Shannon. Bush thought the voder might be useful as an element in his hypothetical Memex, an idea for a memory-extending desk that was to be used as an early information management system. Though never built, the Memex inspired the hypertext theory of Theodore Nelson, now used throughout the Internet.

The voder synthesized speech by creating the electronic equivalent of a human vocal tract, and was advertised as the first machine to create speech that did not rely on the recording technologies of film and phonograph. The voder hearkened back to the Bell Lab's original research in phonetics and speech, and their goal of helping the deaf, as Dudley's instrument had the possibility of giving those who were mute the ability to talk.

In outward appearance, the voder looked like an organ. There had been previous attempts at building organs that could mimic human speech, first by Christian Gottlieb Kratzenstein, a Russian scientist teaching physiology in Copenhagen in 1773. He built an instrument that made sounds approximate to human vowels using resonance tubes connected to organ pipes. In the 1780s, Wolfgang von Kempelen, who was notorious for his bogus chess playing automaton, created an Acoustic-Mechanical Speech Machine that relied on bellows. His machine could make consonant sounds as well as vowel sounds. In the early nineteenth century, Sir Charles Wheatstone, who had been a telegraphic innovator, improved upon the design of von Kempelen, with a version that could spit out a couple of words.

Homer Dudley's speech organ, the voder, had white keys that produced vowels, black keys that produced consonants, and a foot pedal could be used to control the pitch. On the inside, oscillators and noise generators provided a source of pitched tone and hiss. A ten-band resonator filter controlled by the keyboard converted these tones and hisses into the vowels, consonants, and inflections. Another set of extra keys allowed the operator to make the plosive sounds such as "p" and "d," as well as affricate sounds such as "j" in "jaw" and "ch" in "cheese." Only after months of practice with this difficult machine could a trained operator produce something recognizable as speech.

In order to prepare for its world premier, twenty-four telephone operators, called "voderettes," trained on the voder for an entire year. At the World Fair, Mrs. Helen Harper, who was noted for her skill on the voder, led a group of voderette's in demonstrations of the machine, where people from the crowd could come up and ask the operator to make the voder say something.

The voder would later go on to inspire the work and research of Werner-Meyer Eppler, one of the teachers and mentors of Karlheinz Stockhausen, who saw in it the ability to create electronic music.

Homer Dudley had great success in his aim of reducing phone line bandwidth with the vocoder. It could chop up voice frequencies into ten bands at 300 hertz, to the very low end of what could be used on a phone call, a significant reduction from the high end of 3.4 kilohertz. Yet it never got used for that purpose. The large size of the equipment was impractical to install in homes and offices across the country, even if it could create more channels on the phone lines. For a time Dudley worked at marketing the vocoder to Hollywood for use in audio special effects. It never made much of an impact there either, as other voice changing devices, such as the Sonovox, had already begun to be used in radio jingles and in cartoons. However, before it was discovered by musicians, Homer Dudley's tool for voice compression was successfully put into service during America's efforts in World War II. There it became a key part of the SIGSALY encryption program, employed to protect the radio transmissions of high ranking officials.

SIGSALY: Cryptography, Turntables, and Muzak

By the 1940s, radio and telephone had revolutionized communications in the industrialized world. Its strategic use in warfare, as a way to transmit information and intelligence, was not lost on military personnel. Yet these signals needed to be protected. Especially with radio, it was easy to become an undetected eavesdropper wherever signals could be received. After World War I, the first devices to attempt secure transmission of voice were developed. These were substitution devices that inverted frequencies in a way similar to how alphabetic substitution codes swapped one letter for another. High frequencies were substituted for low frequencies and vice versa. It was easy to make a device that could do this, but the system was also easy to break.

In 1931, Bell Telephone Laboratories developed the A-3 scrambler that was used by Roosevelt and Churchill when they talked on the phone. Although the United States was still not at war in 1941, they were aiding the Allies with materiel, and secure calls had to be made. The A-3 worked by dividing the voice frequency into five subbands. Each of these was inverted. On top of the measure of substitution, it also used transposition, with the voice shifted from one subband to another every twenty seconds.

The security of the A-3 was eventually compromised by Germans based at a radio post in South Holland who had been intercepting the British prime minister's telephone calls in 1941. Aware that their system was insecure, the situation was becoming intolerable for the Allies. Tensions had been mounting between the US and Japan, and a "warning of war" had come in late November, but no one knew when or where. On December 7,

1941, American cryptanalysts were hard at work breaking a long ciphertext intercepted from Tokyo to the Japanese embassy in Washington D.C. The team of codebreaker William Friedman, of the Signals Intelligence Service, had broken Japanese diplomatic ciphers before, but this naval cipher, scrambled in fourteen parts, was impenetrable.

One part they did tease out: the Japanese embassy had been instructed to destroy their cryptography equipment and meet with the US Secretary of War, suggesting that armed conflict seemed imminent. Admiral Stark, Chief of Naval Operations, needed to send Admiral Kimmel at Pearl Harbor another message to reinforce the warning of war. Stark knew the A-3 scrambler was no longer secure, so he chose to send his message by radio. However, atmospheric propagation conditions over the Pacific stalled his message in its tracks. The next option was to send a message by undersea cable via Western Union. Unfortunately, by the time it arrived the battleships of the Pacific Fleet had already been bombed and the Japanese planes were flying back to their bases.

The need for secure transmissions became even more paramount as the United States entered the war. In 1942, the Army contracted Bell Labs to assist with the communication problem and create "indestructible speech" that could withstand attempts at code breaking. From this effort, the revolutionary twelve-channel SIGSALY system was born. SIGSALY was not an acronym, just a codename for the project. To create SIGSALY, workers sifted through over eighty patents in the general area of voice security, settling on Homer Dudley's vocoder to form the basis of the system. For SIGSALY, a twelve-channel vocoder was used. Ten of the channels measured the power of the voice signal in the audio portion of the frequency spectrum where most talking occurs, while two channels were devoted to "pitch" information and whether or not unvoiced (hiss) energy was present. The vocoder enciphered the speech as it went out over phone or radio. In order to be deciphered at each end of the conversation, an audio crypto-key was needed. This came in the form of vinyl records.

From the standpoint of music history, it is interesting to note, as Dave Tompkins did in his book *How to Wreck a Nice Beach: The Vocoder from WWII to Hip-Hop*, that the SIGSALY system employed two-turntables and a microphone. The classified name for this vinyl part of the operation was SIGGRUV, while the vinyl records were produced by the Muzak Corporation, a company famous for the creation of elevator music. The sounds on these records weren't aimed at soothing weekend shoppers or people sitting in waiting rooms, but rather contained random white noise, like channel 3 on an old television set. The noise was created by the output of very large mercury-rectifier tubes that were four inches in diameter and over a foot high. These generated wide band thermal noise that was sampled every twenty milliseconds. The samples were then quantized into six levels of equal probability. The level information was converted into channels of a frequency-shift keyed audio tone signal recorded onto a vinyl master. From the master, only three copies of a key segment were ever made. If any SIGGRUV vinyl still exists, and for security reasons they shouldn't, those grooves are critically rare.

It had to be insured that no pattern could be detected, so the records had to be random noise. Even if the equipment had somehow been duplicated by the Axis powers, the communications would still be uncompromised, as they still required the matching vinyl crypto key at each terminal. This made the transportation of these records, via armored truck, the most secure since Edison invented the phonograph. Just as the masters were destroyed after making three keys, each vinyl key was only ever to be played once, as operators were instructed to burn them after playing. The official instruction read, "The used project record should be cut-up and placed in an oven and reduced to a plastic biscuit of 'Vinylite.'" As another precaution against the grooves falling into enemy hands, the turntables themselves had a self-destruct mechanism built into them that could be activated in case one of the terminals was compromised. Thinking of all this adds a fascinating new dimension to the idea of a DJ battle.

Synchronizing turntables at two different terminals across the globe was another technical hurdle Bell Labs overcame, which was no small feat given that if a needle jumped or the system went out of sync, only garbled speech was heard. At the agreed upon time, say 1200 GMT, operators listened for the click of the phonograph being cued to the first groove. The turntables were started by releasing a clutch for the synchronous motor that kept the turntable running at a precise speed. Fine adjustments were made using fifty hertz phase shifters (Helmholtz coils) to account for delays in transmission time. The operators would listen for full quieting of the audio as synchronization was established. Oscilloscopes and shortwave receivers were also used to keep systems locked to international time.

A complete SIGSALY system contained about forty racks of heavy equipment comprising vacuum tubes, relays, synchronous motors, turntables, and custom made electromechanical equipment. In the pre-transistor era, all of this gear required a heavy load of power, so cooling systems were also required to keep it all from getting fried from the heat. The average weight of a set up was about fifty-five tons.

The system passed the inspection of Alan Turing, "the father of modern computer science," if not his test. He had been briefly involved with the project on the British side, just as Claude

Shannon had been in America. On July 15, 1943, the inaugural connection was established between the Pentagon and a room in the basement below Selfridges Department Store in London. Eventually a total of twelve SIGSALY encipherment terminals were established, including locations in Paris, Algiers, Manila, Guam, Australia, and one on a barge that ended up in the Tokyo Bay. In the year 1945 alone, the system trafficked millions of words between the Allies.

To keep all of this operational, a special division of the Army Signal Corp was set up: the 805th Signal Service Company. Training commenced in a school set up by Bell Labs and soldiers were sent to various locations, requiring security clearances and a firm grasp on the cutting edge technology they were tasked to operate and maintain. For every eight hours of operation, the SIGSALY systems required sixteen hours of maintenance and calibration.

In putting the system together, eight remarkable engineering "firsts" were achieved. A review conducted by the Institute of Electronic and Electrical Engineers in 1983 lists them as follows:

1. The first realization of enciphered telephony
2. The first quantized speech transmission
3. The first transmission of speech by Pulse Code Modulation (PCM)
4. The first use of companded PCM
5. The first examples of multilevel Frequency Shift Keying (FSK)
6. The first useful realization of speech bandwidth compression
7. The first use of FSK - FDM (Frequency Shift Keying-Frequency Division Multiplex) as a viable transmission method over a fading medium

8. The first use of a multilevel "eye pattern" to adjust the sampling intervals (a new, and important, instrumentation technique)

SIGSALY left the world with a rich inheritance that spans developments in cryptology and digital communications, a legacy that ticks away in the background of history every time someone hits the decks with two turntables and a microphone.

5: MUSIC BY NUMBER: THE MATHEMATIKOI

AUDREY: Speech Recognition

The other side of the speech synthesis coin is speech recognition. Without the ability for machines to both recognize speech and synthesize it, the wunderkinds of today's consumer electronics, Siri and Alexa, would not be possible. With speech synthesis and recognition in place, humans can now speak, and sometimes yell and cuss, at the wide range of interconnected devices that will speak back to us.

In 1952, over half a century before Alexa could respond to a voiced question of where to find the best noodle shop in town, AUDREY was first on the scene. She derived her name from her special power: Automatic Digit Recognition. She was a collection of circuits capable of perceiving numbers spoken into an ordinary telephone. Due to the technological limits of the time, she could only recognize the spoken numbers of "zero" through "nine." When the digits were uttered into a mic on the handset, AUDREY would respond by illuminating a corresponding bulb on the front panel of the device. It sounds simple today, but this marvel was only achieved after overcoming steep technical hurdles.

As with other aspects of telephony, early efforts at developing speech recognition were built on the backbone of research into acoustics and phonetics. Phonemes, the distinct units of sound within a language, articulation, and the exact ways vocal cords vibrate to produce vowels and consonants were all studied in detail. With vowels, for example, the vocal cords vibrate in a manner that excites the vocal tract. The air becomes a medium of propagation featuring a natural mode of resonance called a "formant," a concentration of acoustic energy centered around a particular speech frequency. There are several formants, about one in every 1,000 hertz band, with each corresponding to a resonance in the vocal tract. The vowels are distinguished from each other by a difference in overtones, with each vowel having three formants or overtone pitches.

In 1952, the Bell Lab researchers who developed AUDREY, S. Balashek, R. Biddulph, and K. H. Davis, created a system that measured or estimated the formant frequencies of vowels within the spoken digits of zero through nine. The goal was to provide automated direct dialing by voice, with other services to be added as the system developed. These automated calls needed to work all across the country and beyond, for people with thick regional accents and distinct ways of speaking. It needed to work for someone without training, who just picked up the phone and told it what number to call.

One of the obstacles they faced was to craft a system capable of recognizing the same word when said with these subtle variations. For instance, the spoken digit "seven" is subject to slight differences when uttered multiple times by even the same person. Duration, intonation, quality, volume, and timing all change the sound of the word with each utterance. To recognize speech amidst all these variables, AUDREY focused on the sound parts within the words that have the most minimal variation. In this way the machine did not need to have an exactly spoken match. Speech researcher Roberto Pieraccini put it this way, saying there is less variety "across different repetitions of the same sounds and words than across different repetitions of different sounds and words."

The closest matches came from the formants. The information that humans require to distinguish speech sounds can be represented in a spectrogram by peaks in the amplitude and frequency spectrum. Thus, AUDREY would locate the formant in the spectrum of each utterance and use that to make a match.

In order to work, AUDREY required that there be pauses between spoken numbers, as she couldn't isolate or separate individual words when said in a continuous string. In practice, it wasn't just anyone who could walk up to AUDREY and start talking to her. She could only recognize the speech of people who had been trained to talk in a controlled manner. She required these designated talkers to produce the specific formants, or else she might not recognize a digit. For each speaker, the reference patterns that were drawn electronically and stored within her memory had to be fine tuned. Despite all the limitations around her use, the researchers proved that building a machine capable of recognizing human speech wasn't a pipe dream.

Like many of the other trailblazing machines already discussed, AUDREY was expensive because she was state of the art and entirely analog. The six-foot high relay rack occupied with all her vacuum-tube circuitry required a lot of upkeep, and she drew a lot of power. The invention never really went anywhere as a tool in Ma Bell's vast monopoly, but it was a stepping stone towards the dynamic voice recognition systems taken for granted today, and other researchers soon followed down the path AUDREY opened.

Later in the 1950s, RCA Laboratories built another system that recognized ten syllables from one particular speaker, while the MIT Lincoln Lab built a speech recognizer that could decipher ten syllables independent of who was saying them. The Japanese were also hard at work, with Radio Research Lab in Tokyo producing a vowel recognizer and Kyoto University creating a machine that recognized phonemes. The latter breakthrough became important because it introduced the process of segmenting speech into different parts for machine analysis. With the growth of machine learning in computer science, vast datasets of speech have been able to be compiled over time, leading to the eventual development of continuous speech recognition systems.

The fact that a machine can be made to decipher strange human vocalizations at all is scientific wonder. AUDREY was a kind of magic and her name deserves pride of place in the history of speech recognition.

Taking It to the MAX: MUSIC I-III

The pure research conducted at Bell Laboratories was widely diffused, and the electronic music systems that arose out of those investigations were incidental and secondary byproducts. The voder and vocoder explored earlier were just the first of these.

American computer music pioneer Max Mathews led the way at Bell Labs with his innovative MUSIC program, the first computer program for sound generation that was broadly available, at least to people with access to places like Bell Labs. Born in Nebraska, in 1926, Mathews fell in love with music when he learned to play violin as a kid, a love affair that deepened when he started listening to records as a teenager. In 1950, Mathews found his second love, computers, as a student at MIT, where he got a side job in the school's Dynamic Analysis and Control Laboratory. Utilizing the lab's analog computer, the child of a differential analyzer developed during World War II by Vannevar Bush, Mathews studied the missile control systems for anti-aircraft missiles as the deep freeze of the Cold War settled over America. On this computer he also learned to work with servomechanisms and do numerical integration, skills that served him for the rest of his life.

Mathews got his doctorate in electrical engineering in 1954 and landed a job at Bell Labs doing acoustic research. Max Mathews' boss was a guy named John R. Pierce, who had done essential work on Pulse Code Modulation (PCM) with Claude Shannon and Bernard M. Oliver. PCM is a way to transform analog signals into a digital binary, the same technique used to encode the SIGSALY transmissions, and the three men received patents and together published the paper "The Philosophy of PCM" in 1948. Pierce had done extensive work in the fields of radio communication, microwave technology, and psychoacoustics—not to mention writing science fiction under the name J. J. Coupling—and had coined the word "transistor." He went on to do work on the Telstar 1 satellite, write a number of popular works on topics like information theory, and work with Max Mathews on computer music.

At Bell, Mathews now had access to state of the art digital computers such as the IBM 650, used to develop speech compression and coding for their undersea long-distance lines to Europe. Mathews figured out a way to digitize and manipulate sound on the computer, convert it to audio, and play it back using PCM. It was a convoluted process that put to work a theorem developed by Harry Nyquist and Claude Shannon.

The Nyquist-Shannon theorem has come to be known as the sampling theorem. It is fundamental to converting signals from analog-to-digital and back again from digital-to-analog. Much modern technology works with its application ticking away in the background. When converting a signal, there is always danger that essential information may be lost. Nyquist and Shannon did the difficult theoretical work that made those losses minimal. The theorem shows the minimum sampling rate that a continuous-time signal needs to be uniformly sampled from, so that the original signal can be remade by the samples alone. Their theorem has also become the backbone of Digital Signal Processing (DSP). It is now routine for music producers and sound technicians in the film industry to use DSP effects on all manner of recordings. The whole idea of sampling in music, an indispensable element in the hip-hop, electronic, and pop music of today, can be thought of as going back to Shannon and his theories.

In the beginning, when Max Mathews was applying the theorem to his speech coding research, it wasn't so seamless, and all the work had to be done using tape and mainframe computers. Digital tape recorders were a key component of this work. IBM had created these special devices, but they built their own at Bell Labs to use for voice sampling. These weren't recordings of audio, but information, so they would then take this tape up to IBM in New York, where the team used their computers to write programs for converting the tape into audio. The audio on this tape was not speech yet. It had to be reconverted again with the equipment back in New Jersey to make it speak. The tape recorder was used to speed up the computer's output, using the Nyquist frequency, this being the highest frequency which can be coded at a given sampling rate to reconstruct a signal. The individual samples were twelve bits, holding more information than an analog tape recorder would fit. The computer would run all night, and back at Bell Labs the tape would be played back in real time. The work heralded a new phase of their research into sound and speech, and allowed Bell Labs to become the pioneers of computer music.

These days people don't have to work so hard. Somebody running Pro Tools in a home studio built in their attic can now use DSP technology that was once costly and in the hands of just a few.

None of the work Mathews had done so far had been used yet for musical purposes, even though he and his supervisor were avid music aficionados, with Mathews playing the violin and Pierce playing the piano. One night when the two were at a concert together, during the intermission, Pierce said, "Max, if you can get sound out of a computer, could you write a program to synthesize music on a computer?" Mathews was up for the challenge and Pierce encouraged him to take time off his speech coding research to focus on writing a music program.

The result was MUSIC I, which Max Mathews created in 1957. That same year Lejaren Hiller and Leonard Isaacson had managed to compose music with computers, creating their "Iliac Suite" by carefully programming a computer to write a score for a string quartet by using randomly generated numbers, which were then either accepted or rejected according to a set of rules based in music theory. In essence Hiller and Isaacson were playing a kind of game with their machine, with the results of their game then played by human musicians.

MUSIC I was a different kind of beast, as the computer would also generate the sound of the composition. The compositional approach was markedly different from Hiller and Isaacson's. First, the notes were selected and then written out as MUSIC subprograms. The computer was then instructed which subprograms to use and combine. All these numbers were then spat back out by the computer and transformed into audio, and the results recorded to tape.

MUSIC I was a beginning. It gave a programmer control over the pitch, scale, and notes of a sound, but only one voice and timbre. Newman Guttman, a linguist and psychoacoustics researcher who worked at Bell, was the first to write a piece with the composing software. His "The Silver Scale" lasts only seventeen seconds and sounds like some random notes from a lo-fi video game. Later the same year Guttman and Mathews created "Pitch Variations," which is a minute long. These weren't meant to be polished pieces of art, but rather experiments to show it could be done. Refinement came later.

It would have been easy to give up on the project at this point, because the first results were not very listenable. Instead they persevered, guided on by the theories of Claude Shannon. In principle, computer music recording and playback could create any sound that the human ear could hear. This is untrue for all traditional instruments. Computer music could even create timbres as of yet unheard. There were many hurdles to tackle, but the idea of creating these new sounds gave Max Mathews and those he collaborated with the gumption to keep going. They received further encouragement from musicians and composers who saw the potential of the computer. These included Edgard Varèse, Vladimir Ussachevsky, and Milton Babbitt, all of whom were ecstatic about the door being opened to new musical possibilities.

The next iterations of the program attempted to do more. MUSIC II was not very different from the original program, but it did have the ability to use four voices, which allowed for polyphony, and it came with more synthesis algorithms that gave the programmer greater controls for varying timbre. Mathews overcame major obstacles with MUSIC III, which was written as a block diagram compiler. Code written in block diagram form is very similar to the blueprint schematic of a circuit, where each block is a collection of codes that do different things. For MUSIC III, Mathews had blocks for a generalized oscillator, filters, mixers, and noise generators, which when connected together dramatically expanded the possibilities for composing music.

A Bicycle Built for Two

Speech synthesis confers a number of benefits to its users. It allows individuals with impaired eyesight to operate radios and computers. For those who cannot speak and may also have trouble using sign language, speech units such as the device employed by Stephen Hawking allow a person to communicate in ways unthinkable a century ago. For these people, speech synthesizers play an integral role in adding quality to their day to day lives.

Beyond these specialized uses, one of the ways everyone can share in the joy of speech synthesis is through the medium of music. Bell Labs was again at the forefront of making this happen, through the work of John Larry Kelly Jr.

Kelly was a close colleague of Claude Shannon, and is today most often remembered for his Kelly criterion, utilized by gamblers and financial investors alike. Born in Texas in 1923, Kelly was a young man by the time World War II was underway, spending four years as a pilot in the Air Force. When he got out of the service, he got his PhD in physics, with a specialty in investigating "second order elastic properties of various materials," which led him to work in the oil industry. The oil company he worked for had no real use for him, however, as his employer was a successful wildcatter with a nose for sniffing out oil. When he was offered a job at Bell Labs, he headed out to New Jersey. Still not thirty when he went to the lab, he was distinguished by his young age, his Texas drawl, and his intensity and intelligence. Kelly and his wife, Myldred, were avid card players and he liked to gamble. While at Bell Labs, one of the things he was working on was developing compression schemes for television, which put him into contact with Claude Shannon and the field of information theory. Kelly was able to make connections between TV, gambling, and information theory that led him to the development of the Kelly criterion.

Kelly had become convinced that gamblers who had a little "inside information" could make greater returns on their bets by the liberal application of Shannon's equations. Shannon liked Kelly's idea and pushed him to publish his findings in the *Bell System Technical Journal*. He wanted to call his paper, "Information Theory and Gambling," but AT&T didn't like the title, as they did provide wire services for bookies and others involved in organized crime, and didn't want others to think they were actively helping criminals earn more money. Instead, his findings got published under the less exciting title, "A New Interpretation of Information Rate."

The application of the Kelly criterion began in earnest in the 1960s, when Ed Thorp, a student at MIT, mentioned to Shannon he had developed a card counting system for blackjack, who in turn gave him an even bigger meal ticket by telling Thorp about Kelly's theory of probability. Thorp successfully used the strategy to place optimal bets, and later used it to make money as a hedge fund manager, after which Wall Street too took an interest in the Kelly criterion.

After this work, Kelly became head of the Bell Labs information and coding department. It was in this capacity that he worked with Carol C. Lochbaum and Louis Gertsman to develop a new speech synthesis program, which they demonstrated by programming the computer to sing the well-known standard, "Daisy Bell (Bicycle Built for Two)," composed by Harry Dacre in 1892.

Gertsman was a neuropsychologist who ended up specializing in the fields of aphasia research and linguistics, which in the early 1950s led him to become an associate in a project on the perception of speech at Haskins Laboratories. At Bell Labs he was also the director of a project on computer simulation of speech, working on the computer that would become an inspiration for HAL 9000 in *2001: A Space Odyssey*.

The computer Kelly and Lochbaum coaxed into talking and singing was the IBM 704, first introduced in 1954, with 140 units sold by 1960. The programming languages LISP and FORTRAN were first written for this large machine that used vacuum tube logic circuitry. In 1961 Kelly wrote a vocoder program for the 704, and demonstrated it by having it sing "Daisy Bell."

Kelly and Lochbaum's work was the first digital physical-modeling example of speech synthesis. They had used speech-vowel data from a paper by Gunnar Fant, "The Acoustic Theory of Speech Production." Fant obtained his vocal tract data from X-rays of vowels spoken in Russian, not English, but it was close enough to guide them to what they needed to do. The year after their demonstration, Kelly and Lochbaum published an article on their work, which came to be known as the Kelly-Lochbaum Vocal Tract model.

Lovely as the a cappella computer was, it was deemed in need of instrumental accompaniment. For this part of the song, they sought out the expertise of fellow Bell Labs employee Max Mathews, who programmed the accompanying 8-bit melody portion of "Daisy Bell" using the IBM 7090.

The IBM 7090 was the transistorized version of the 709 vacuum tube mainframe. The 7090 series was designed for "large-scale scientific and technological applications." With a price tag of close to $3 million, the first of the 7090s was installed in late 1959. Adjusted for inflation, the price today would be a whopping $23 million. Besides its musical capabilities, the 7090's other accomplishments include controlling the Gemini and Mercury space flights, running the Air Force's Ballistic Missile Early Warning System up until the 1980s, and being used by Daniel Shanks and John Wrench to calculate the first 100,000 digits of pi. Yet none of the above applications compare, at least in this author's mind, to the beauty of the IBM 704 joining forces with the IBM 7090 on the song "Daisy Bell."

HAL 9000 still gets most of the credit for this electronic version of "Daisy Bell." As it so happens, *2001: A Space Odyssey* author Arthur C. Clarke was visiting his friend and colleague John Peirce at Bell Labs when John Larry Kelly was making his demonstrations of speech synthesis with the IBM 704. He was so fascinated by witnessing this computational marvel that he wrote "Daisy Bell" into his screenplay for Stanley Kubrick's 1968 film adaption, in which HAL signs the song in the middle of the machine's climactic mental breakdown.

Kelly was an intense person who liked to shoot guns and play cards in his off hours. He also chain smoked, sometimes going through six packs of cigarettes a day, dying from a stroke in 1965 at the age of forty-one while walking down the street in Manhattan. After his death, the line of speech synthesis research based on vocal-tract analog models that he had pursued began to fade out. Spectral models had taken the lead for their ease of application in communication technology, as linear predictive coding (discussed below) came into the ascendancy. With these techniques, high fidelity audio transmission could be achieved at low bit rates, which was embraced by telephone companies and, as a byproduct, presented new applications in the world of music.

In 1969, Gertsman helped work on the proto-ambient album, *Environments 1*, with Irv Teibel. At Bell Labs, on an IBM 360 computer, he helped process a sound loop of crashing waves, recorded at Brighton Beach, New York. The subsequent recording was used on the A side of *Environments 1* and titled "The Psychologically Ultimate Seashore." The recording was designed to be played at various speeds, from the standard rotation of 45 RPM, all the way down to 16 or 23 RPM, making it one of the longest LPs ever recorded. Released nearly a decade before Brian Eno's genre-coining "ambient" albums, *Environments 1* introduced the idea of listening to natural environmental field recordings for the purpose of relaxation and meditation to a wide public, helping pave the way for the ambient and "new-age" records that would follow.

The field of speech research seems to connect everything, from gambling, to Kubrick's famous films, to stock trading, to notorious court cases, as was the case when Louis Gertsman went on to become an expert in voiceprints. A voiceprint is supposed to be like a fingerprint, and a spectrogram of someone's voice can show an individual's distinctive speech patterns. To identify a voiceprint, graphs of taped voice samples and spectrograms are compared. In 1973 Gertsman testified in the trial accusing New Orleans District Attorney Jim Garrison—best known for his investigations into the John F. Kennedy assassination—of bribery charges. Gerstman claimed that the tape that was supposed to be Garrison speaking had actually been spliced together and was a fraudulent fabrication, leading to Garrison's acquittal.

Music from Mathematics

Between 1960 and 1962, Bell Labs joined forces with the Decca record label to introduce computer music to the public with the first phonograph record of this novel mode of electronic music. The result was the 1962 release *Music from Mathematics: Played by the IBM 7090 Computer and Digital to Sound Transducer.* The tracks featured a number of selections composed by the musical engineers at Bell Labs, showcasing what had been accomplished using versions of MUSIC I-III, but it was the rendition of "Bicycle Built for Two" that remains the most famous song on the album. The record also featured the original Mathews composition "Numerology," hearkening back to the Pythagorean origins of music and math.

Six compositions from engineer John R. Pierce were included, with titles like "Variations in Timbre and Attack" and "Five Against Seven-Random Canon." Pierce, who had a long-term interest in the confluence of art, science, and technology, would later go on to become one of the independent co-discoverers of a non-octave musical scale, later named the Bohlen–Pierce scale. The album was rounded out by James Tenney's "Noise Study," Orlando Gibbons' "Fantasia," and computerized takes on familiar titles such as "Frere Jacques" and "Joy to the World."

The sounds on the album have a preponderance of the basic waveforms used in synthesized music: the square, sawtooth, and triangle waves. Square waves at low frequencies tend to sound like clicks, while higher frequencies sound like buzzing. The jagged drop-off of sound created by the sawtooth gives it an even punchier buzz than the square tone, while a triangle wave is somewhere between the buzz of a square and the smooth waves produced by sine generators. Vibrato and various types of noise rounded out the still limited palette, giving the record something of a clinical sound, but not an unpleasant one. It is rather what you might expect for these proof of concept pieces where the familiar is mixed with the alien, and the worldly with the otherworldly. Even these first steps were enough to fire off the imagination of listeners.

Don Slepian was one such young listener. His father, Bell Labs mathematician David Slepian, was a colleague of Claude Shannon who made fundamental contributions to information theory and coding. His work contributed to the Joint Photographic Experts Group image compression scheme, leading to the pervasive JPEG image format that underlies the modern internet. During the early phase of MUSIC, David Slepian led a small project in aleatoric composition started by Mathews. Each mathematician on the team was shown four bars of written music and asked to contribute two bars to follow. This was repeated for each participant.

Although the results were unfocused and didn't end up being used on the *Music from Mathematics* album, David Slepian did bring home a test pressing of the record in 1961, and his son listened to it so often that he eventually wore out the grooves. Don Slepian later wrote of his listening experience, "As an eight year old child I was passionately in love with this record, playing it over and over again every day as I absorbed every musical drop. It was as if the music I always heard in my head suddenly became magically real. I wanted to make computer music myself but I was only eight." In time this dream became a reality when he became a technician for Max Mathews at the labs, and later an artist in residence.

MUSIC IV-V

Max Mathews completed MUSIC IV in 1963, with the assistance of Joan Miller. That same year his influential paper, "The Digital Computer as a Musical Instrument," was published in *Science* magazine. In it he wrote, "there are no theoretical limitations to the performance of the computer as a source for musical sounds, in contrast to the performance of ordinary instruments." Those interested in the young field of electronic music had been on an ongoing quest for new timbres, and now the computer was being shown to be a capable tool for the creation of completely original sound worlds.

New to MUSIC IV was a graphical interface that could work as a front-end to the software being written for the mainframe computers by Max Mathews. Note values were written on a cathode ray tube computer monitor and transformed into sound by digital synthesis, bypassing the often protracted procedures for programming and inputting the data. Once stored on the memory, the inputs could be modified in various ways or duplicated, saving precious time on the machine.

MUSIC V was written in 1967. Mathews was careful to give artists working with his MUSIC programs complete compositional control. While timbre and instrument voice were not defined, the way that the blocks or unit generators were connected created a voice. He also made it possible to program a musical score as a computer file. Time, notes, pitch, duration could all be programmed. Other computer programmers interested in MUSIC followed Mathews' lead and wrote numerous variations for different computers and programming languages.

In 1964, French composer and physicist Jean-Claude Risset came to Bell Labs and started to work with Mathews and John Pierce, experimenting with sound synthesis and psychoacoustics. Risset had come over with a degree in music and obtained a degree in physics during his time at Bell, his intense interest in each informing the other. Risset started using MUSIC IV to synthesize the sound of brass instruments and do work on sound processing. Made using MUSIC V in 1968, his "Computer Suite from Little Boy" is considered one of the first significant pieces of music realized entirely on computer.

Risset wrote the music for "Computer Suite from Little Boy" to accompany a play by Pierre Halet, which revolves around a pilot who is tortured by reliving the bombing of Hiroshima in a dream that leads to a psychological break. Risset showed off his adaptation of Roger Shepard's forever-rising chromatic scale, reversing it into an ever-descending glissando that mimics the pilot's psychological descent. Risset also makes use of other sound illusions he was able to realize, such as sounds that appear to decrease in speed but are really increasing, and rhythms that seem to slow down as the tempo increases. With Risset encouraging his passion for computer music, young Don Slepian became fascinated by this barber pole effect that became known as the Shepard-Risset glissando.

Through synthesis, Risset had learned how to "compose sounds themselves, just as one composes a chord; composing spectra helps to impart to timbre a functional role and not only a coloristic one, " as he wrote in the *Computer Music Journal*. Using spectral analysis of chords, he takes the harmonic arpeggios that open the piece and, instead of hearing them simultaneously, scans the harmonics of the notes forming the chord from top to bottom in succession. Later in the piece, Risset does a trick where he uses chords "that slowly glide with a constant frequency difference between the components." These are also transposed on a scale that matches the intervals between the chord components. According to Risset, he went "further than composing with sound, I exploited the possibility of composing the timbres themselves." In doing so, he fulfilled Mathews' dream of creating a kind of music that went beyond the limitations inherent in traditional instruments.

6: Variations For Speech

After Kelly and Lochbaum's work with computer programmed speech synthesis, the vocoder reached a forking path. One path went straight towards music, the other went deeper into the worlds of telephony and· radio. The path towards greater musical use was taken up by Robert Moog and Wendy Carlos, while the telephonic branch manifested in the subfield of linear predictive coding, developed independently by Fumitada Itakura and Bishnu J. Atal, which eventually got repurposed for music too, in a roundabout way, when in got into the hands composer of Paul Lansky, and later, by way of the Speak & Spell learning toy, Q. Reed Ghazala.

Robert Moog, Wendy Carlos, and A Clockwork Orange

Before Robert Moog gave the world his namesake synthesizer, he cut his teeth making theremins. Born and raised in New York City, his father was an engineer for Consolidated Edison and a ham radio operator, so electronics were a part of household life. Though his parents had paid for his harp lessons, he liked to hang out with his dad in the workroom, where all the tinkering happened. Upon learning of the theremin, he became fascinated with the instrument, and at age fourteen, in 1949, he built his own based on plans he found in an issue of *Electronics World* magazine. Four years later, Moog had produced his own theremin design and published his own article on it in *Radio and Television News.* That same year he started his company, RA Moog, to sell theremin kits through the mail. American composer Raymond Scott happened to be one of his customers, rewiring Moog's theremin to create another early electronic instrument, the clavivox.

Moog's dedication to electronic music continued to grow, and he pursued degrees in electrical engineering, eventually earning his PhD from Cornell. Before he had put that feather in his cap, the Moog synthesizer was already a reality. Moog's synth became a hit with musicians because he had made something relatively affordable, compared to room-sized beasts such as the RCA Mark II, which required hundreds of vacuum tubes. The Moog was made up of modules connected by patch cords, and though large, was not nearly as cumbersome as its predecessors. This created greater interest from working musicians, not just the academic types who had access to places such as the Columbia-Princeton Electronic Music Studio.

Wendy Carlos was one of the musicians who helped turn on the switch of Moog's synth. A child of Rhode Island, Carlos went to Brown University to study physics and music before going on to Columbia in New York City, where Moog had gotten his Masters of Science. The two met at the school's annual audio engineering show and began working together. In addition to giving Moog a lot of nuanced advice from a player's point of view, Carlos got her own Moog synthesizer, soon becoming one of its first champions.

After releasing his synth to the public, Moog started getting interested in vocoders. In the past vocoders had been built using vacuum tube technology, but in 1968 Moog built one of the first solid state, or semiconductor, vocoders. This was followed by a ten-band device inspired by Homer Dudley's original designs, and worked with his modular synthesizer. The carrier signal came from a Moog synth, while the modulator was the input from the microphone. The brilliant application of this instrument made its debut appearance in Stanley Kubrick's 1971 film, *A Clockwork Orange*, where the vocoder sang the vocal part from the fourth movement of Beethoven's Ninth Symphony, the section titled "March from a Clockwork Orange" on the soundtrack. This was one of the first recordings made with a vocoder, and it's a fascinating coincidence that two early uses of speech synthesis for music ended up in films made by Kubrick. The song "Timesteps," an original piece written by Carlos, also features on the soundtrack. She had originally intended to include it as a mere introduction to the vocoder for those who might consider themselves "timid listeners," but Kubrick surprised Carlos by its inclusion in his dystopian masterpiece.

Coming down the road in 1974 was the classic album *Autobahn*, by the German krautrockers Kraftwerk. This was the first commercial success for the group, following their worthwhile but highly experimental first three albums. Kraftwerk's contribution in the popularization of electronic music remains huge, and their use of the EMS Vocoder 1000 was cutting edge at the time their records first came out while still sounding fresh today. Besides using other commercial gear such as the EMS Synthi AKS, Minimoog, and ARP Odyssey, Kraftwerk were dedicated homebrewers of their own instruments. It has been reported that one of their home-built instruments, the Robovox, was patented by band member Florian Schneider in 1990. This instrument was used to synthesize singing voices in real time, and can supposedly be heard on their song "The Robots" from their 1991 album *The Mix*.

Three years later, in 1977, British rock band Electric Light Orchestra released their hit seventh album, *Out of the Blue*, which represents the first major pop album to make comprehensive use of the vocoder. Singles like "Mr. Blue Sky" and "Sweet Talkin' Woman" are songs that toggle the happy switches in the brain, and ELO would continue their love affair with the vocoder across their subsequent recordings.

During the 1980s, as the music technology became increasingly available to the public, the vocoder became ever-more popular among musicians across the spectrum, including the fledgling genre of hip-hop. While the vocoder was enjoying great success in the entertainment industry, its use in telecommunications was also still ticking away, albeit a bit quieter, in the background.

Linear Predictive Coding

The elements of linear predictive coding (LPC) were based on the work of Norbert Wiener, a computer scientist who became "the father of cybernetics" back in the 1940s. Wiener developed a mathematical theory for calculating the optimal filters for finding signals in noise. Claude Shannon quickly followed Wiener with his breakthrough paper, "A Mathematical Theory of Communication," which included a general theory of coding. With new mathematical tools in hand, researchers started exploring predictive coding. "Linear prediction" is a form of signal estimation, and it was soon applied to speech analysis.

In signal processing, communications, and related fields, the term "coding" generally means putting a signal into a format where it will be easier to handle a given task. A coding scheme like Morse Code, for instance, is when an encoder takes the signal and puts it into a new format. The decoder takes it out of its new format and puts it back into the old one.

The "predictive" aspect of coding has been used for countless scientific theories and engineering techniques. What they all have in common is that they predict future observations based on past observations. Joined together, the term "predictive coding" was coined by information theorist Peter Elias, who wrote two papers on the subject in 1955.

In LPC, samples from a signal are predicted using a linear function from previous samples. In math, a linear function has either one or two variables without exponents, or it is a function that graphs to the straight line. The error between a predicted sample and the actual sample is also transmitted along with the coefficient, or number used to multiply a variable. This works with speech because the samples from nearby correspond to each other to a high degree. The error is also transmitted because if the prediction is good, the error will be small and take up less bandwidth. In this sense, LPC becomes a type of compression based on source codes.

Towards the end of the 1960s, both Fumitada Itakura and the duo of Bishnu S. Atal and Manfred Schroeder independently discovered, as in the case of the telegraph and telephone, the elements of LPC. Later, American composer Paul Lansky applied it to make delightful music exploring the spectrum between music and speech and when Texas Instruments created their Speak & Spell learning toy that employed LPC, it got creatively repurposed for music by Q. Reed Ghazala.

Fumitada Itakura: From Whistlers to Formants

Japanese scientist Fumitada Itakura was interested in math and radio from an early age. An amateur radio operator in his youth, his elementary school happened to be just a mile from the radio laboratory at Nagoya University, where his father knew some of the professors, allowing him to visit it and ask questions.

As an undergraduate, Itakura became interested in the theoretical side of math and started to learn about probability theories such as stochastic processes, or sequences of random variables. As he extended his ability ever further, he eventually became involved in the mathematical aspect of signal processing. His research paper for his bachelor in electrical communication was on the statistical analysis of whistlers, a very low frequency electromagnetic radio wave produced by lightning and capable of being heard on radio receivers. To study it, Itakura built a bank of analog filters to do the signal processing, adding digital circuits to try and find patterns in the time-frequency of the whistlers. It wasn't easy work, but he persevered.

In analyzing the whistler signal, he had to filter out a lot of the other noisy material that comes in from the magneto-ionosphere, requiring him to use band-pass filters and the sound spectrogram that had originally been designed for speech analysis. This eventually led to further work with statistics and audio. When he went to graduate school he studied applied mathematics under professor Kanehisa Udagawa, at whose lab he researched pattern recognition, starting a project to recognize handwritten characters in 1963. When professor Udagawa died of a heart attack, Itakura had to find someone else to study under to continue his course, leading him to work at the Nippon Telegraph and Telephone Corporation (NTT).

Dr. Shuzo Saito had been a graduate of Nagoya University and was looking for someone to research speech recognition with. Saito's friend, professor Teruo Fukumura, suggested Itakura. Fukumura began privately teaching Itakura the basic principles of speech using Gunnar Fant's "Acoustic Theory of Speech Production." Once on board, Itakura started making sound spectrograms of his voice speaking vowels. His voice was high and husky, so it didn't make as clean of a spectrogram as it would have with a more common voice. In this, there was a hidden gift. Itakura realized if they could do good analysis on a signal like his voice, which had more random characteristics, they could do even better when analyzing regular speech. From this point, he applied statistics to speech classification, based on a paper he had read by J. Hajek.

Dr. Saito suggested that Itakura look for practical results based on his theory, so he started working with a vocoder and got some initial results on his idea. Saito then suggested he look at pitch detection, as vocoders often had trouble recognizing voices because of their poor ability in this area. Itakura conceived of a new method of pitch detection that used an inverse filter and oscillation, integrating the linear predictive analysis to create a new vocoder system. In late 1967 he succeeded in synthesizing speech from the vocoder and brought the results to Dr. Saito. Fumitada Itakura would center his work around vocoding for the rest of his career.

Of the many modes in which speech is produced, the way vowels sound is very important. This is because it relies on the periodic opening and closing of the vocal cords, where air from the lungs gets converted into a wideband signal filled with harmonics containing many properties. This signal resonates the vocal cavities before leaving the mouth where the final sounds are shaped.

In vocoding, this speech signal gets analyzed, the signal of the formants estimated and removed in a process called inverse filtering. The remaining sounds, called the buzz, are also estimated. The signal that remains after the buzz is subtracted is called the residue. Numbers which represent the formants, the buzz and the residue, can be stored or transmitted elsewhere. The speech is then synthesized through a reversal of the original stripping process, wherein the parameters of the buzz and residue are used to create a signal, and the information stripped from the formants is used to create a new filter.

This method of vocoding took speech apart at the millisecond level of time and put it back together on the other end, saving a ton of bandwidth. This huge technical feat by Fumitada Itakura enabled telecommunications companies to fit five calls onto the same channel where an unsynthesized voice call would take up one. Itakura was invited to work at Bell Labs from 1973 to 1975 in the Acoustics Research Department.

Mafred Schroeder and Bishnu S. Atal: Coding and Decoding

German mathematician Manfred Schroeder was born in 1926, and like Fumitada Itakura he held an early fascination with radio. He came of age during World War II and showed his technical aptitude when he built a secret radio transmitter, which really freaked out his parents. Transmitting radio was risky business at the time, because it was the province of spies and people who wanted to communicate outside the country, so when Schroeder saw members of the army or SS outside his house with radio direction finding equipment it spooked him and he shut off his transmitter for a month. He also liked to listen to the BBC and the American Forces Network transmitting from England. Many people had been sent to concentration camps just for listening to foreign stations and spreading news to others. The Nazi powers attempted to keep tight control on all information going in and out of the country, even going so far as to manufacture a special radio receiver, the Volksempfänger (People's Radio), that could only receive approved German stations under the directorship of Joseph Goebbels.

Schroeder excelled at school and was often ahead of his teachers. During the war he was drafted to a radar team to track incoming aircraft flights, gaining extensive experience with the technology. Like Itakura, Schroeder was also a math fanatic, and when he did go to university, he always took extra math classes on the side of his physics work. He had been fascinated by crypto math, loading up on function theory and probability classes. At Physics Colloquium at Göttingen University, Schroeder was in the class when Warner Meyer-Eppler, Karlheinz Stockhausen's later mentor, started to talk to his students about Shannon's information theory, mentioning that there were many research opportunities available at Bell Labs. After applying two times, Schroeder got a job offer from Bell in 1954, based on previous work he had done experimenting with microwaves, and quickly emigrated to the United States.

At Bell Labs, many researchers were left alone to pursue the dictates of their own curiosity and see where it took them. Though Bell wanted Schroeder to continue his research with microwaves, he thought he'd switch gears and get into the study of speech instead. For two years he worked on speech synthesizers, but didn't have much luck in getting them to sound good. He then turned his attention to speakers and room acoustics.

When John Pierce urged him to use Dudley's vocoding principles to send high fidelity voice calls over the phone system, Schroeder found himself facing the same issue that Itakura had: the problem of pitch. Part of the issue was extracting the fundamental frequencies from telephone lines not known for superb sound quality in the first place. As Schroeder investigated, he realized he could take the baseband signal, or those frequencies that have not been modulated, and distort it non-linearly to generate frequencies that the vocoder would then give the right amplitude. This breakthrough became known as "voice-excited vocoding," and the speech that came out of the other end was the most human sounding of any speech synthesis up to that point.

In 1961, Schroeder hired Dr. Bishnu S. Atal to work with him at Bell Labs. Born in India in 1933, Atal studied physics at the University of Lucknow and received his degree in electrical communications engineering from the Institute of Science in Bangalore in 1955, before coming to America to earn his PhD at the Brooklyn Polytechnic Institute. He returned to his home country to lecture on acoustics from 1957 to 1960 before being lured back to the US by Schroeder.

In 1967 Schroeder and Atal were pacing around the lab, conversing about the need to do more with vocoder speech quality. Schroeder's work on pitch had improved the quality of vocoding, but it wasn't yet what it could be. What they needed to do, they realized, was to code speech so that no errors were present. As they talked, the idea of predictive coding was born.

Schroeder and Atal realized that as speech became encoded, they could predict the next samples of speech based on what had just come before. The prediction would be compared with the actual speech. Alongside this, even errors in data, known as residuals, would be transmitted, so nothing would be lost. In decoding, the same algorithm was used to reconstruct the speech on the other end of the transmission. Schroeder and Atal called this "adaptive predictive coding," with the name later changed to linear predictive coding. The quality of speech was as good as that which came out of Schroeder's voice-excited vocoder. They wrote a paper on the subject for the *Bell System Journal* and presented it at a conference in 1967, the same year Itakura succeeded with his technique.

Since the 1970s, most of the technology around speech synthesis and coding has been focused on linear predictive coding, which remains the most widely used form. When it first came out, the National Security Administration (NSA) was among the first to get their paws on it because LPC can be used to securely transmit a digitized and encrypted voice sent over a narrow channel. The early example of this is Navajo I, a telephone built into a briefcase used by government agents, of which about 110 were produced in the early 1980s.

LPC has become essential for cell phones, and is a standard protocol for cellular networks like the Global System for Mobile Communications (GSM). LPC is also used in Voice Over IP, or VoIP, platforms such as Skype and Zoom. Vocoding technology is also utilized in the Digital Mobile Radio (DMR) units that have gained popularity among amateur radio enthusiasts around the world.

Godfrey Winham and Kenneth Steiglitz: Variations for Speech

In 1962, a computer center was opened at the brand new engineering quadrangle of Princeton University. The school was home to some forward thinking engineers, and like their colleagues at Bell Labs, some of them were serious music heads. Godfrey Winham was an English-born music theorist, composer, and computer programmer. Winham had come to Princeton at the behest of Milton Babbitt when the two had met in Salzburg. Babbitt was taken by Winham's brilliance and persuaded him to come study with him and Roger Sessions. He stayed and got his undergraduate, his graduate, and received the school's first doctorate in composition, and then became a professor.

Kenneth Steiglitz was an engineer and jazz aficionado who had joined the faculty at Princeton in 1963 as an assistant professor. Steiglitz has fond memories of dissecting discarded radios as a child, like many others who got bit with the engineering bug. As he familiarized himself with the campus, he heard some sounds coming from a door. It happened to be Winham, who was working with another music professor, James K. Randall. The two were trying to get a digital tape drive to work with a set of loudspeakers, and had been working on synthesizing music with the new IBM computers, and MUSIC software given to them by Max Mathews. Mathews had also supplied a digital to analog converter, but it was well used, and had soon stopped functioning. Steiglitz thought he might be able to help these musicians, who had no official technical training, as they wrestled with the technology.

Godfrey Winham had become fascinated with the possibilities for applying computers to music and this became the main area of his focus until his untimely death in 1975. He was only able to put in twelve years of focused work on composing with computers, but in that time he transferred his enthusiasm to the students who took his graduate seminar in computer synthesis. Steiglitz was able to provide specialized technical support and the two became collaborators. Winham found a more than able guide in Steiglitz, who helped steer his friend through a convoluted maze of circuitry and programming.

Winham developed his own variation of MUSIC IV, dubbing it MUSIC 4B, and his variation started to spread to other universities. Another version came out when Hubert Howe rewrote the program for use in the Fortran computer language. This became MUSIC4BF in 1966, the same year Paul Lansky arrived at Princeton and enrolled in Godfrey Winham's class.

Paul Lansky: Mild und Leise

Paul Lansky was born into a musical family in 1944 New York, where his father worked for Capitol Records. Growing up familiar with music studios, he recalls being fascinated by the drum set up at one of Count Basie's recording sessions. Lansky went to the High School of Music and Art in Manhattan, and got his undergraduate degree in Queens College, where he studied French horn composition. From there he went to Princeton to work on his graduate degree. Lansky's teachers included George Perle and Milton Babbitt, and from these figures he was schooled in the methods of twelve tone music and serialism. The RCA Mark II Synthesizer had become the centerpiece of the Columbia-Princeton Electronic Music Studio, and between the studio and the computer center, young composers were encouraged to start working with electronic music. To this end Paul Lansky enrolled in a graduate seminar in computer synthesis taught by Godfrey Winham.

The class was very exciting for Lansky who thought Winham was a "young genius." With help from Steiglitz, Winham started to tackle the many problems around computer music at that early stage. They started by taking spoken recordings and manipulating them to sound octaves higher or lower. Fellow computer music composer Charles Dodge was among the attendees of this seminar, along with a cast of other characters.

The computer in the quadrangle had been upgraded to an IBM 7094, and was free for the music students to use, yet without the digital-to-analog converter, they couldn't hear their music. As the new and younger student, Paul Lansky was often tasked with driving their digital tapes forty miles into New Jersey to Bell Labs for conversion under Mathews hospitality.

Working under the auspice of Babbitt's complex mathematical take on serialism, Lansky put in long hours at the computer programming a twelve-tone piece, but when he finally completed it and listened back, he thought it sounded like garbage, and even Babbitt wouldn't give him much in the way of useful commentary. Frustrated, Lansky gave up on using the computer for composition for four years. Switching gears, he started to work with an iconoclastic "twelve tone tonality" system developed by his other teacher, George Perle. Lansky used the method to write a number of instrumental pieces, of which only "Modal Fantasy" for solo piano has survived. In the meantime, his hope for the possibilities of the computer as a musical tool were revived when he heard J.K. Randall's, "Lyric Variations for Violin and Computer," which he found inspiring and imaginative.

Meanwhile, Barry Vercoe had written a new version of MUSIC called Music360. This program was installed on Princeton's next computer, the IBM 36/91. They now had their own digital-to-analog converters again, and Paul Lansky gave computer composition another shot. "Mild und Leise" was the result. A proper old school effort, as far as the computing method is concerned, the procedure involved using a series of punch cards, and learning the mechanics of the system was as much a part of the work as writing the score. Incorporating the harmonic language from Perle's system into the work, which gives it a strong emotional resonance, Lansky felt as if he had finally accomplished something. "It took a year to complete and I sweated bullets over every note," he later wrote. Showing its long-lasting appeal, a four-chord sequence of "Mild und Leise" was later sampled by Radiohead in their song "Idioteque" on their *Kid A* album after Jonny Greenwood stumbled across the track when crate digging at a used record shop.

Speech Songs

For all the triumph Lansky felt in completing "Mild und Leise," he felt there were some issues with it, and he went in search of new sounds. He had really liked the "Speech Songs" of Charles Dodge and thought these might offer a new direction for him to go. "Speech Songs" had in part come out of the work Godfrey Winham had started doing with Steiglitz on synthesizing the human voice. They had taken spoken word recordings and had learned how to manipulate them to sound octaves higher or lower using LPC. In their experiments they had written a Fortran programming language subroutine to do the math, but there was a problem. Winham had Hodgkin's lymphoma and his life was cut short at the age of forty. Before Winham passed away, he had given a set of cards with the LPC infused music program on it to Charles Dodge.

Charles Dodge was one of the composers who had frequented Bell Labs to use their digital to analog converters to be able listen to his work. On his trips there, he became mesmerized by the fascinating sounds of the speech research going on down the hall, and often thought it was more interesting than the sounds he'd created using computers. In the early seventies, Dodge had the opportunity to create some new works at Bell Labs with the box of cards for the LPC program given to him by Winham. He also worked with Joseph Olive who was a leading researcher in the area of text-to-speech and one of those people who had both an intense mathematical mind and a strong interest in music.

With help from Olive, and some poems written by his friend Mark Strand, Dodge went about creating the album *Speech Songs*. Of the project he later wrote:

> *I'd never been able to write very effective vocal music and here was an opportunity to make music with words. I was really attracted to that. It wasn't singing in the usual sense. It was making music out of the nature of speech itself. With the early speech-synthesis computers, you could do two things: you could make the voice go faster or slower than the speed in which it was recorded at the same pitch or you could shift the pitch independent of the speech rhythm. That was a kind of transformation that you couldn't make in the usual way of making tape music. It was fascinating to put my hands on two ways of modifying sound that were completely, newly available.*

To synthesize the electronic voices for the poems, Dodge used what he called "speech-by-analysis," where only words that he had put into the computer before, using an analog-to-digital converter, could be synthesized. The recorded speech was analyzed by the computer to pull out the various parameters from the spoken word in short segments. Then speech could be recreated by the artificial voice using the same parameters as had been analyzed. This was another artistic use in applying the Nyquist-Shannon theorem. For musical purposes, though, those parameters could be altered to change aspects of the sound, such as shifting the pitch contour of a phrase or word into a melodic line. Changing the speed without altering the pitch was another possibility, as was adjusting the formants and resonance.

The poems in *Speech Songs* are humorous and surrealistic, and the way the artificial voice reads them adds to the effect that the words are being read by a drunken chorus of robots. Dodge was specifically interested in humor because, as he wrote in the liner notes, "Laughter at new music concerts, especially in New York, is rare these days." He was delighted when audience members laughed at his creation. For a type of music that is so often cerebral and conceptual, it's good when some chuckles can be had.

Another piece on the album, "The Story of Our Lives," also used techniques of speech synthesis. In this case, instead of replacing the recorded human with an artificial voice, they changed the program so that it took from a bank of sixty-four sine tones that glissandoed at different rates. To create the effect of more than one voice being heard at a time, the different voices were mixed together on the digital computer. The end result is like listening to a bunch of robots at a beatnik poetry reading, only these digitized poets are taking big hits off a helium tank before each recitation of verse. Voices rise for a time, and then go back down as the helium leaves the lungs. Another voice in the piece sounds like a sinister criminal from a sci-fi thriller, only they are talking about mundane yet existential matters.

Idle Chatter

After Godfrey Winham's death, Ken Steiglitz and Paul Lansky both aimed to carry on the work of their friend. LPC became the keystone in Lansky's compositional quest for new sounds, and he asked Steiglitz to write a program using LPC where he could shift the formants, allowing him to make a man's voice into a woman's voice, or to make a woodwind sound like a viola. To do this, the timbre was altered by changing the speed. This led to "Artifice," Lansky's first piece using LPC in 1976.

LPC had its problems too, but Lansky thought "it was exciting to imagine being free of the binding of pitch, rhythm and timbre." In 1978 he began his next experiment with LPC with "Six Fantasies on a Poem by Thomas Campion." Lansky's wife, Hannah McKay, reads the poem, while a variety of processing and filtering methods are used to alter and transform the reading in fabulous ways. In his notes on the recording of "Six Fantasies," Lansky writes about how it has become common to view speech and song as distinct categories, countering, "they are more usefully thought of as occupying opposite ends of a spectrum, encompassing a wealth of musical potential. This fact has certainly not been lost on musicians: sprechstimme, melodrama, recitative, rap, blues, etc., are all evidence that it is a lively domain."

A composer as well as a poet in his own time, Thomas Campion became an archetype, emblematic of the "musical spectrum spanned by speech and song." Embedded within his 1602 treatise, *Observations in the Art of English Poesie*, the poem Lansky used was "Rose cheekt Lawra." In it, Campion offered his attempt at a quantitative model for English poetry, where meter is determined by the quantity of vowels, rather than by rhythm, as was done in ancient Greek and Latin poetry. Lansky describes the poem as such:

> *[a] wonderful, free-wheeling spin about the vowel box. It is almost as if he is playing vowels the way one would play a musical instrument, jumping here and there, dancing around with dazzling invention and brilliance, carefully balancing repetition and variation. The poem itself is about Petrarch's beloved Laura, whose beauty expresses an implicit and heavenly music, in contrast to the imperfect, all too explicit earthly music we must resign ourselves to make. This seemed to be an appropriate metaphor for the piece.*

Lansky continued to explore the continuum between speech and song with his humorous and logically named pieces, "Idle Chatter," "just_more_idle_chatter," and, "Notjustmoreidlechatter." Though clearly connected by theme, they are not a suite, but independent works. "Idle Chatter," from 1985, continues with the use of his wife as vocalist and the IBM 3081 as the means of transforming her voice, and again he uses a mix of LPC, stochastic mixing, and granular synthesis, with a bit of help from the computer music language Cmix, which Lansky wrote and developed himself. If you like the sound of speaking in tongues and always wanted to hear what it sounded like at the Tower of Babel, these recordings are your best opportunity. Of the project, Lansky wrote,

> *The incoherent babble of Idle Chatter is really a pretext to create a complicated piece in which you think you can 'parse the data', but are constantly surprised and confused. The texture is*

> *designed to make it seem as if the words, rhythms and harmonies are understandable, but what results, I think, is a musical surface with a lot of places around which your ear can dance while you vainly try to figure out what is going on. In the end I hope a good time is had by all (and that your ears learn to enjoy dancing).*

People had a strong reaction to the piece, and in response to their reaction, Lanksy wrote, "just_more_idle_chatter" in 1987. He gave the digital background singers more of a role in the piece, but the words still only approach intelligibility and never really reach a stage where the listener can comprehend what is being said. The next saw his "stubborn refusal to let a good idea alone" with the realization of "Notjustmoreidlechatter." Here again the chatter almost becomes something that can be discerned as a word before slipping back down into the primordial soup of linguistic babble.

"I wanted to use the computer more as an aural camera," Lansky said. His compositions are like polaroid snapshots, photographic masterpieces steeped in the chemical baths of a dark room revealing captured light. Lansky's recordings have captured human and machine speaking together in a harmonic dialogue.

Over time, though Lansky wrote many more computer music pieces and settings for traditional instrumentation, he couldn't let the words just be. For the pieces on his album, *Alphabet Book*, he conducted further investigations in a magisterial reflection on the building blocks of thought: alphanumerics, the letters and numbers, that allow for communication, the building up of knowledge, and contemplation.

Speak and Spelling with Q. Reed Ghazala

A tenth order derivative of LPC was used in the popular 1980s educational toy, Speak & Spell. These became popular to hack by experimental musicians in a process known as circuit bending, where the toy is taken apart and the connections re-soldered to make sounds not intended by the manufacturer.

The process of circuit-bending was pioneered by Qubais Reed Ghazala, who cultivated a whole suite of instrument types from the detritus of junk sound making toys he found in thrift shops and other places. The Speak & Spell became a favorite for Reed to bend, and in his expert hands became what he called an Incantor.

Speak & Spells were a product of Texas Instruments, originating as an outgrowth of the company's research into speech synthesis, and first introduced to the public in 1978 at the Consumer Electronics Show. This high tech toy consisted of a TMC0280 linear predictive coding speech synthesizer, an alphabetic keyboard, and a receptor slot to receive one of a collection of ROM game library modules.

The Speak & Spell used trademark Solid State Speech technology that stored full words in a solid state memory format that was similar to the way calculators from the same era stored numbers. The expansion modules could be inserted through the battery receptacle to load new libraries and games. It was the first educational toy to reproduce speech without relying on tape or phonograph recording, and words could be punched in and spoken in a way similar to how Texas Instrument calculators could solve a math problem. The original intention of the unit, as advertised, was as a tool for helping kids around age seven and up to learn the spelling and pronunciation of difficult and commonly misspelled words.

The word libraries were created from recordings of professional speakers brought in by Texas Instruments to utter and say the words. Once the voices were captured they needed to be further processed to fit the limited memory of the ROMs. Once processed the words often needed further editing because of the sharp reduction and cut of the original data rate. Information had been lost and noise had been introduced into the system as the words were compressed. Some of the recorded words had become completely unintelligible.

All the hard work required by the technicians and engineers at Texas Instruments got bent to other purposes when Ghazala got his hands on a Speak & Spell. A Speak & Spell out of the box is already musical, and after one of the units terminals gets cross-wired and additional electronic components such as potentiometers get installed, after the normal functioning is completely disrupted, it becomes an "incantor," capable of incanting from the basic parameters of letters, words, phonemes, and vowels. Incantors end up speaking in tongues and talking in alien languages.

These modifications tend to overwhelm the unit's keyboard switch matrix and trigger an effect known in the field of electronics as key jamming or ghosting. This is something that happened on older matrix keyboards when three keys were pressed together at once, making a fourth keypress to be erroneously registered by the keyboard controller. The circuit glitches and weird sounds emerge. In this manner a glitch became a feature.

Once rewired, all of Ghazala's instruments get the beauty treatment. They are repainted and made into true one-of-a-kind art objects, the equivalent of a luthier applying the final stains and varnish to violin or guitar. Reed has built many of his sculptural experimental instruments for interesting musicians and bands, including Tom Waits, Peter Gabriel, Pat Mastalotto from King Crimson, Faust, Towa Tei from Deee-Lite, and the band Blur.

Circuit-bent sounds in popular music are difficult to document, but the unaltered Speak & Spell's have shown up on several albums and releases. Depeche Mode notably named their 1981 debut synth-pop album *Speak & Spell* after the toy. Beck used a Speak & Spell on his 1994 song "In a Cold Ass Fashion," employing it to give texture to his iconoclastic melding of styles, while in the same year Pink Floyd used it on their meditative song "Keep Talking." In the year 2000 the alternative pop rock band Self released their album *Gizmodgery*, recorded using only toy instruments, many of them circuit bent. An Incantor seems to be able to be heard on the track "Chameleon" with its hip-hop influenced driving beat and eruptions of glitched out noise.

Ghazala's own song, "Silence the Tongues of Prophecy" is a classic example of his Incantor in use, uttering alien languages in an electronic glossolalia punctuated by guttural low-end intonations and textures evocative of extraterrestrial worlds. Ghazala has insisted that circuit-bending is a folk art, something able to be learned by anyone with the time and desire to do it. As such, circuit-bent Speak & Spells have often mostly been found out in the wilds of underground music culture, played in small venues for small crowds, and heard on self-released albums or small labels devoted to electronica and noise, new folk musics for the electronic age.

The Telephone Game

The musical ramifications of Homer Dudley's vocoder and the subsequent research into speech and its synthesis had a wide ranging effect on the later development of electronic music. It can be looked at as a sequence of transformations akin to the telephone game played by kids. One kid says something and it gets repeated by another, only slightly altered, but by the time it has been whispered from mouth to ear several times, inevitable changes have been wrought. The inventor of Auto-Tune, Andy Hildebrand, used many of the elements for speech analysis, such as LPC and formant analysis, to create his pitch correcting tool. The sound of all that research is now being used in all kinds of popular music, from Cher to Daft Punk to Bon Iver, with Auto-Tune especially having been adopted in the hip-hop genre, where it has become something of a sonic signature for artists like T-Pain and Travis Scott.

Auto-Tune, vocoders, and computer music are only one suite of tools developed by the marriage of telecommunications technology and music. To find the others, we have to look at the way radio stations instigated the birth of electronic music studios.

PART III

WE ALSO HAVE SOUND-HOUSES

7: HEARING INNER SOUNDS: THE AKOUSMATIKOI

Arts of Noise

In our machine dominated age, there is hardly any escape from noise. Even in the most remote wilderness outpost, planes will fly overhead to disrupt the sound of birds and the wind in the trees. In the city, noise has become so much a part of the background we have to tune in to the noise in order to notice it, because we've become adept at tuning it out. Roaring motors, the incessant hum of the computer fan, vibrations from the refrigerator, metal grinding at the industrial factory down the street, the roar of traffic on I-75, the beep of a truck backing up... These and many countless other noises are part of our daily soundscape.

Throughout human history, musicians have sought to mimic the sounds around them. The gentle drone of the tanpura—a stringed instrument that accompanies sitar, flute, and voice in classical Indian music—was said to mimic the gentle murmur of the rivers and streams. Should it be a surprise then, that in the nineteenth and twentieth century, musicians and composers started to mimic the sounds of the machines around them?

In bluegrass and jazz there are a whole slew of songs that copied the entrancing rhythms of trains. As more and more machines filled up the cities, is it any wonder that the beginnings of a new genre of music—noise music—started to emerge? Could it be any coincidence that as acoustic and sound technology progressed, our music making practices also came to be dominated by machines? And just what is music anyway? There are many definitions from across the span of time and human culture. Each definition has been made to fit the type, style, and particular practice or praxis of music.

In his 1913 manifesto, *The Art of Noises,* the Italian Futurist thinker Luigi Russolo argues that the human ear has become accustomed to the speed, energy, and noise of the urban industrial soundscape. In reaction to those new conditions he thought there should be a new approach to composition and musical instrumentation. He traced the history of Western music back to Greek musical theory, which was based on the mathematical tetrachord of Pythagoras and did not allow for harmony. This changed during the Middle Ages, first with the invention of plainchant in Christian monastic communities, which employed the modal system to work out the melody, but did not denote the exact pitches or intervals that needed to be sung. Nonetheless, these relative notes embodied the first revival of musical notation after knowledge of the ancient Greek system was lost. In the late ninth century, plainsong began to evolve into organum, which led to the development of polyphony. Until then, the concept of the musical chord did not exist as such.

Luigi Russolo thought that the chord was the "complete sound." He noted that in history chords developed slowly over time, first moving from the "consonant triad to the consistent and complicated dissonances that characterize contemporary music." He pointed out that early music tried to create sounds that were sweet and pure, and then it evolved to become more and more complex. By the time of Schoenberg and the twelve-tone revolution of serial music, musicians sought to create new and more dissonant chords. These dissonant chords brought music ever closer to his idea of "noise-sound."

As part of the Futurist movement, Russolo thought a new sonic palette was also required. He proposed that electronics and other technology would allow Futurist musicians to create new timbres unavailable in the traditional orchestra. His imperative was to "break out of this limited circle of sound and conquer the infinite variety of noise-sounds." This would be done with technology that would allow us to manipulate noises in ways that never could have been done with earlier instruments.

Russolo wasn't the only one thinking of the aesthetics of noise, or seeking new definitions of music. French Modernist composer Edgard Varèse said that "music is organized sound." It was a statement he used as a guidepost for his aesthetic vision of "sound as living matter" and of "musical space as open rather than bounded." Varèse thought that "to stubbornly conditioned ears, anything new in music has always been called noise," and he posed the question, "what is music but organized noises?" An open view of music allows new elements to come into the development of musical traditions, where a bound view would try to keep out those things that did not fit the preexisting definition.

Out of this current of noise music initiated in part by Russolo and Varèse, a new class of musician would emerge: the musician of sounds. These musicians would be most at home in the studio, the radiophonic laboratories, and sound-houses where music could be assembled from electrons, taken apart, reconfigured, and put back together again in alternate forms.

Expressions of Zaar

Ancient Egypt gave birth to many things. In Egypt the word *Khem* was used to describe the fertile floodplains surrounding the Nile river. Their highly evolved religious beliefs, their mastery of magic, and their practices around mummification all gave rise to an extensive body of knowledge centered around a goal of spiritual immortality. This body of knowledge would include aspects of what became alchemy, and later chemistry. In the twentieth century, Egypt also gave birth to the first experiments in manipulating recorded sounds, leading to the creation of a new kind of music—a kind of alchemy itself—all while searching for the properties of an inner sound.

Born in 1921, into an affluent Coptic Christian household hailing from the Upper Egyptian province of Asyut, Halim El-Dabh was eleven when his family moved to the Cairo suburb of Heliopolis. Following in the footsteps of his father, El-Dabh went to school for agricultural engineering. On the side, he studied music, and received recognition within his country for his compositions and piano technique.

El-Dabh had always been fascinated with the traditional and folk music of his homeland, and later the world. The same year they moved to Heliopolis, his brother took him to a conference on Arab music where, for the first time, he heard music that had been recorded on a wire, the predecessor to tape recording.

This eventually led to his becoming an ethnomusicologist, but first it led him to Middle East Radio in Cairo. Borrowing one of the wire recorders, he went out into the streets and recorded an ancient Zaar exorcism ceremony. He thought that by processing the sound in the radio station's studios, he could investigate the "inner sounds" of the ceremony. To find this inner sound, he later recalled:

> *I just started playing around with the equipment at the station, including reverberation, echo chambers, voltage controls, and a re-recording room that had movable walls to create different kinds and amounts of reverb… I concentrated on those high tones that reverberated and had different beats and clashes, and started eliminating the fundamental tones, isolating the high overtones so that in the finished recording, the voices are not really recognizable any more, only the high overtones, with their beats and clashes, may be heard.*

El-Dabh was the first to take raw recordings and transform them into a new work. The results are a ghostly audio manifestation of a ritualistic space, conjuring up a kind of shadow world where the original sounds can only be heard as prismatic reflections, prefiguring the subgenre of dark ambient by decades. Presented in 1944, his finished twenty minute piece, "The Expression of Zaar," later released as "Wire Recorder Piece," predated Pierre Schaefer's experiments with musique concrète by several years.

After graduating from Cairo University in 1945, El-Dabh was invited to study at the University of New Mexico, receiving scholarships to Brandeis University and the New England Conservatory of Music as well. He settled with his family in New Jersey and found himself ensconced in the avantgarde music scene, hanging out with the likes of Edgard Varèse, John Cage, and Henry Cowell.

In the 1950s he would make even more adventurous electronic recordings. These connections, and a friendship forged with Otto Luening and Vladimir Ussachevsky led him to create significant works at the Columbia-Princeton Electronic Music Studios as part of his musical destiny as detailed in Chapter Nine. Later his interest in ethnomusicology took him to many nations in the continent of Africa and elsewhere around the world to collect field recordings for his own music and lectures. He later served the Smithsonian Institute as a consultant. In Egypt, at the Great Pyramid of Giza, his music has been played every evening since 1961 as part of the Sound and Light show projected on the ancient sacred site.

Pierre Schaeffer: Musician of Sounds

A French experimental audio researcher who combined his work in the field of radio communications with a love for electro-acoustics, Pierre Schaeffer developed his theory and practice of "musique concrète" during the 1930s and '40s. Musique concrète was a practical application of Russolo's idea of "noise-sound" and the exploration of expanded timbres made possible through then new studio techniques. It was also a way of making music according to Varèse's "organized sound" definition, distinct from previous methods by being the first type of music completely dependent on recording and broadcast studios.

In musique concrète, sounds are sampled and modified through the application of audio effects and tape manipulation techniques, then reassembled into a form of montage or collage. It can feature any sounds derived from recordings of musical instruments, the human voice, sounds created in the studio, or field recordings of natural and man-made environments. Because Schaeffer was the first to use and develop studio music making methods, which were soon adopted by the likes of Karlheinz Stockhausen and the founders of the BBC Radiophonic Workshop, he is considered a pioneer of electronic music, and by extension one of the most influential musicians of the twentieth century. The recording and sampling techniques he used, such as recording sounds on to an acetate record to then further in manipulate in the studio, and later his exploration of tape with loops, became part of the standard operating procedures used in recording studios around the world, earning him the title "Musician of Sounds."

Born in 1910, Pierre Schaeffer had a wide variety of interests during his eighty-five years on this planet. He worked variously across the fields of composing, writing, broadcasting, engineering, and as a musicologist and acoustician. His work was innovative in science and art. It was after World War II that he developed musique concrète, all while continuing to write essays, short novels, biographies, and pieces for the radio. Much of his writing was geared towards the philosophy and theory of music, which he later demonstrated in his compositions.

It is interesting to think of the influences on Schaeffer as a person. Both his parents were musicians—his father a violinist and his mother a singer—but they discouraged him from pursuing a career in the arts, instead pushing him into engineering. He studied at the École Polytechnique south of Paris, where he received a diploma in radio broadcasting. Thus, he combined the perspective and approach of an engineer with his inborn musicality to bear on his various activities.

In 1934, Schaeffer got his first telecommunications gig in Strasbourg, before getting married, having his first child, and moving to Paris where he began work at Radiodiffusion Française (now called Radiodiffusion-Télévision Française, or RTF) the following year. As he worked in broadcasting, he started to drift away from his initial interests in telecommunications and towards music. When these two sides met he really found his niche.

After convincing the radio station management of the alternate possibilities inherent in the audio and broadcast equipment, as well as the possibility of using records and phonographs as a means for making new music, he started to experiment. He would record sounds to phonographs and then speed them up, slow them down, play them backwards, or run them through other audio processing devices, mixing his newly minted sounds together. While all this is just par for the course in today's studios, it was the bleeding edge of innovation at the time.

With these techniques mastered, Schaeffer started to collaborate with people he met via the RTF. All this experimentation had as a natural outgrowth a style that would lend itself to the avant-garde of the day, producing sounds that challenged the way music had been thought of and heard. Out of their combined engineering acumen, new electronic instruments were created to expand on the possibilities in the audio lab, which eventually became formalized as the Club d'Essai, or Test Club.

Club d'Essai

In 1942, Pierre Schaeffer founded the Studio d'Essai, later dubbed the Club d'Essai at RTF. He was joined in the leadership by famed theater director, producer, actor, and dramatist Jacques Copeau. Before becoming a center of musical activity, the Club was active in the French resistance during World War II. It started as an outgrowth of Schaeffer's radiophonic explorations, but with a focus on being active in the Resistance on French radio. It was even responsible for airing the first broadcasts to a liberated Paris in August 1944.

It was at the Club where many of Schaeffer's ideas were put to the test. After the war, Schaeffer had written a paper that discussed questions about how sound recording augments the perception of time, due to the ability to slow down and speed up sounds. The essay showed his grasp of sound manipulation techniques, which were also demonstrated in his early compositions.

In 1948 Schaeffer initiated a formal "research into noises" at the Club d'Essai and on October 5 of that year, he presented the results of his experimentation at a concert in Paris. Known collectively as "Cinq études de bruits" ("Five Studies of Noises"), five works for phonograph were presented. These included the compositions "Etude Aux Chemins de Fer" ("Study of the Railroads"), which utilized the sounds of trains, "Étude aux tourniquets" ("Study of Turnstiles"), featuring sounds of toy tops and percussion, and "Étude pathétique" ("Pathetic Study"), mashing up recordings of sauce pans, boats, vocals, harmonica, and piano. This was the first flowering of the musique concrète style, and from the Club d'Essai another research group was born.

GRM: Groupe de Recherche de Musique Concrète

In 1949 another key figure in the development of musique concrète stepped onto the stage. By the time Pierre Henry met Pierre Schaeffer via Club d'Essai, the twenty-one year-old percussionist and composer had already been experimenting with sounds produced by various objects for six years. Henry was obsessed with the idea of integrating noise into music, and had already studied with the likes of Olivier Messiaen, Nadia Boulanger, and Félix Passerone at the Paris Conservatoire from 1938 to 1948.

For the next nine years Pierre Henry worked at the Club d'Essai studio at RTF. In 1950, he collaborated with Schaeffer on the piece "Symphonie Pour Un Homme Seul." Two years later Henry scored the first musique concrète to appear in a commercial film, *Astrologie Ou Le Miroir De La Vie*, and continued to compose and score for a number of other films and ballets.

Together, the two Pierres were quite a pair, and founded the Groupe de Recherche de Musique Concrète in 1951, later renamed Groupe de Recherches Musicales (GRM). This gave Schaeffer a new studio, which included a tape recorder. Using tape as a recording medium was a significant development for him, as he previously only worked with phonographs and turntables. Using tape sped up the work process, and also added a new dimension: the ability to cut up and splice the tape in new arrangements, something not possible on a phonograph. Schaeffer is generally acknowledged as being the first composer to make music using magnetic tape in this manner.

Eventually Schaeffer had enough experimentation and material under his belt to publish the book *À la Recherche d'une Musique Concrète* (*In Search of a Concrète Music*) in 1952, which was a summation of his working methods up to that point. The design of the GRM studio itself followed the principles Schaeffer had expounded upon in his book. In addition to the regular tape decks used, several new tape manipulation instruments were built just for the studio. These included the three versions of the Phonogène and the Morphophone, all created and built by Schaeffer and Henry's other collaborator at GRM, Jacques Poullin.

Poullin worked with Schaeffer on the design of the various Phonogènes. The Chromatic Phonogène played a tape loop driven by multiple capstans at varied speeds, allowing player control of tape sounds at varying pitches, defined by a small one-octave keyboard, in accordance with the twelve semitones of the tempered scale. The Sliding Phonogène was used to make a continuous tone with the tape speed varied with a control rod. The Phonogène Universal made it possible to transpose the pitch of the sounds on the tape without altering their duration. It also worked in reverse and allowed a composer to alter the duration without changing the pitch. Resembling the regulator later found on VHS video tape recorders, a rotating magnetic head called the "Springer temporal regulator" was used to achieve these possibilities.

The Morphophone, meanwhile, was a tape loop device built in 1954 where the tape was stuck to the edge of a rotating disk with a fifty centimeter diameter. The sound was picked up at different points on the tape by twelve magnetic heads. One recorded, another erased, and ten played back the sounds, which were next passed through a series of bandpass filters (one for each playback head) and amplified.

Schaeffer remained active in other aspects of music and radio throughout the '50s. In 1954, he co-founded Ocora, short for "Office de Coopération Radiophonique," a record label and facility for training broadcast technicians with the purpose of preserving rural soundscapes in Africa through recordings. Doing this kind of work also allowed Schaeffer to make a mark in the world of field recordings and the preservation of traditional music. The training side of the operation helped get people trained to work with the African national broadcasting services.

Before slowing down and finishing his career as a professor at the Paris Conservatoire, Schaeffer's last electronic noise study, the "Etudes aux Objets" ("Study of Objects") was realized in 1959. For Pierre Henry's part, two years after leaving the RTF, he founded the first private electronic studio in France, the Apsone-Cabasse Studio, with French musician and filmmaker Jean Baronnet. Later Henry made a tribute to Schaeffer with his composition "Écho d'Orphée."

Many leading composers of the day, as well as those who later made a name for themselves, were attracted to the equipment, experiments, and energy being diffused from within the GRM studio. Edgard Varèse, Olivier Messiaen, Karlheinz Stockhausen, Pierre Boulez, Jean Barraqué, and Iannis Xenakis were among the luminaries who made music on its machines. Not only were these the leading lights in new music, but the works they created all explored different ways of using the studio as a compositional tool.

Some of the compositions that were made inside the sound-house between 1951 and 1953 alone included the Pierre Boulez studies "Étude I" and "Étude II" (1951), which scratch and scrape along with hard cuts juxtaposed with the hiss of noise. It sounds like the work of an abstract percussionist and is something of a precursor to the harsh noise genre, or the surrealistic splices cut-up by Nurse With Wound. Olivier Messiaen's "Timbres-durées" (1952) plays in a similar vein, with quick transitions punctuated by percussive blasts and electronic fizzes and fades. Soon these composers had discovered the jump cut, made use of in film, where a single continuous take has been edited to make it sound like the listener is moving back and forth in time.

The studio made it possible to rearrange the order of listening, altering ideas about how sounds should be organized in a sequence to create music. In 1952 Karlheinz Stockhausen visited the studio and created "Konkrete Etüde" or "Étude aux Mille Collants" ("Study of a Thousand Splices"), examined in the next chapter. Stockhausen's work at the studio was followed by Pierre Henry's "Le microphone bien tempéré" (1952), with its sliding tones and transmogrified drumming, after which Henry created his celebrated "La voile d'Orphée" (1953), another surreal voyage showcasing the way the medium of tape could be exploited to create soundworlds that before only existed in the imagination. "Orphée 51 ou Toute la Lyre" (1951) was the first opera that employed a musique concrète background tape while a singer performed alone on stage to accompany it. This joint creation by Schaeffer and Henry was overshadowed two years later by "Orphée 53" (1953). With its strange sounds and slow spoken word sections it seems like the perfect kind of music to be diffused over the radio, but it caused a scandal when it was premiered at the Donaueschingen contemporary music festival in Germany, due to its melding of traditional forms with music created in the studio. In 1954 Edgard Varèse and Arthur Honegger used the studio to work on the tape parts of his landmark composition "Déserts" and for "La rivière endormie."

With tape manipulation, musicians had finally caught up with visual artists to create works of collage. Varied sound sources could now be arranged together in ways recalling the disjointed sounds and images of a dream narrative, taking listeners on journeys spanning the whimsical to the sinister, into trippy headspaces never heard before. Their efforts cleared a path for later musical sound collagists such as Frank Zappa, Negativland and People Like Us to go down, just as it gave radio programmers ideas for a type of show that blurred the distinctions between a radio play and avant-garde music.

Inside the Acousmatic Sound Space

Pierre Schaeffer was a man of many energies. One of the musical ideas he devoted himself to, which came out of the practice of making musique concrète, was that of acousmatic sound. The French word "acousmatique" came into the language by way of the the Greek word "akousmatikoi," who were the neophyte students of the philosopher Pythagoras, required to sit in absolute silence while they listened to him deliver his lecture from behind a veil. Pythagoras was said to have used this technique so that his students would concentrate on the sounds being uttered from his mouth, as well as the teachings embedded within his words, rather than his own presence as a teacher. For Schaeffer, acousmatic music was the kind heard from behind the veil of the loudspeaker. When listening to musique concrète and tape music, or recorded music in general, the listener does not see how it was made and played. As in radio, the source is hidden. As Schaeffer wrote in his *Treatise on Musical Objects*:

> *The concealment of the causes does not result from a technical imperfection, nor is it an occasional process of variation: it becomes a precondition, a deliberate placing-in-condition of the subject. It is toward it, then, that the question turns around; "what am I hearing?... What exactly are you hearing" -in the sense that one asks the subject to describe not the external references of the sound it perceives but the perception itself.*

For Schaeffer, acousmatics became an important part of his development of his idea of the sound-object, a kind of primary unit of sonic material, in most cases recorded and distinct from music that was simply notated. This became vital to his sense of aesthetics around ideas of acoustic events being appreciated for their own sake without regard to their original source.

Acousmatic concerts became part and parcel of the experience in new music circles. Listeners would come and sit or stand, just as they would at a regular concert, but with no musicians or instruments in sight. Often the sounds were distributed, not just to one or two speakers, but spatially to multiple speakers surrounding the listener in a practice known as "sound diffusion." The work was often diffused by the composer, or by what Stockhausen called the "sound projectionist," who may or may not be visible on the stage, somewhere in the audience. As acousmatic practice developed, composers often provided in their scores a guideline for the spatialization of the work by interpreters, sometimes called a diffusion score. At its simplest, the diffusion score is a graphic map of the way the sound moves between a loudspeaker arrangement over the course of time. The acousmatic spatialization of sound became an important part of the toolkit for many composers following Schaefer's lead.

François Bayle was a student of both Messiaen and Stockhausen, who went on to have a huge influence on the GRM once Schaeffer stepped away, and he was in charge of the studio from 1966 to 1997. Bayle was also a devotee of acousmatic music, and the main force behind the development of the Acousmonium, a large array of loudspeakers to be used for acousmatic music and sound diffusion.

One of Bayle's first compositions, the "Espaces Inhabitables" from 1967, sounds like the launch of a shuttle into another world, where machines wind up to do their work, and where nature has been transformed. Utilizing numerous field recordings, including those made in a radar dome, zither, and piano, each sound exerts its own gravitational field, creating a sense that space itself has been blurred, and that the relationship between objects has become non-linear. The extensive use of field recordings mixed with electronic sounds became a staple of Bayle's work, which has the capacity of evoking all manner of strange imagery in the listener.

Bayle felt that his purpose as a composer was to stimulate the feeling of motion and vibration of energy that is active within the universe, and within the listener. Describing his approach, he wrote,

> *My purpose was always the same: to compose with only 'images-de-sons' (sound images); to show how, through pure listening in an acousmatic situation, these sound images move like butterflies through audible space and project a colored twinkling on the listener. Out of frame, this is a world proved by itself...*

One of Bayle's major projects for the GRM in the 1970s was a sound diffusion system called the Acousmonium, which grew out of the work of Jacques Poullin's Potentiomètre d'Espace, a system designed in the 1950s to move monophonic sound sources—consisting of a single musical line—across four speakers. Bayle's idea was to create an orchestra of loudspeakers ("un orchestre de haut-parleurs"), and he did this with the help of engineer Jean-Claude Lallemand in 1974. The Acousmonium consisted of between 50 and 100 loudspeakers, depending on the concert and what was needed for the event. Designed specifically to play musique-concrète and acousmatic works, with the added enhancement of loudspeakers placed on the stage and at key points throughout the performance space, a mixer was used to project and diffuse the recorded acousmatic music across the array of speakers, creating various spatial effects. In his own words, Bayle said that the Acousmonium was used to "substitute a momentary classical disposition of sound making, which diffuses the sound from the circumference towards the center of the hall, by a group of sound projectors which form an 'orchestration' of the acoustic image."

An inaugural concert on the system took place on February 14, 1974, at the Espace Pierre Cardin in Paris, featuring a presentation of Bayle's "Expérience Acoustique." As of 2010, the Acousmonium was still pumping out sounds from sixty-four speakers, thirty-five amplifiers, and two mixing consoles.

A grand sign of things to come, the Acousmonium was, in its way, similar to the massive sound systems created for reggae, dub music, and hip-hop in the seventies. Hip-hop can be seen as both musique concrète and acousmatic music in how it relies on the manipulation of turntables, sampled sounds and was originally diffused through sound systems.

A Concrète Legacy

The legacy of Pierre Schaeffer's musique remains concrète. Schaeffer had known of the "noise orchestras" of his predecessor Luigi Russolo, and Edgard Varèse's idea that all music is "organized noise," but took their concepts of noise music further by creating a laboratory of sound for the specific purpose of investigating and playing with noise. He created the basic toolkit that later became the starting point for electronic music experimenters, and he was the original sampler, blazing a trail for hip-hop DJs, turntablists, and sound collagists who would slow records down, speed them up, reverse them, and even press their own records with breaks and samples for creating this new form of popular music.

8: ELECTRIC OSCILLATIONS: THE STUDIO FOR ELECTRONIC MUSIC OF THE WEST GERMAN RADIO

Dr. Friedrich Trautwein and the Radio Experimental Laboratory

The story of the Studio for Electronic Music, the influential facility run by German public broadcaster Westdeutscher Rundfunk (WDR), is linked to the earlier work of two German instrument makers, Dr. Friedrich Trautwein and Harald Bode. Two institutions, the Heinrich Hertz Institute for Research on Oscillations and the Staatlich-akademische Hochschule für Musik, were also critical precursors for the development of the technology around electronic music, in particular the latter's Radio Experimental Lab.

Way back in 1888, Dr. Friedrich Trautwein was born in Würzburg, Germany, and grew to become an engineer with strong musical leanings. After beginning an education in physics, he quit and turned his attention to law with the intention of working for the post office in the capacity of a patent lawyer, protecting intellectual properties around developments in radio technology. When World War I broke out, he became the head of a military radio squadron, an experience that cemented his love for communications technology. After the war ended, he went on to receive a PhD in electrical engineering, and between 1922 and 1924 got two patents under his belt, one for generating musical notes with electrical circuits. Trautwein then went to Berlin in 1923 where he worked at the first German radio station, the Funk-Stunde AG Berlin.

On May 3, 1928, the the Staatlich-akademische Hochschule für Musik (State-Academic University of Music) opened their new department, the Rundfunkversuchstelle (RVS), or the Radio Experimental Lab. One of their goals was researching new directions and possibilities in the development of radio broadcasting. At the time in Germany, much thought was going into the way music was played and heard over the radio. There were many issues around noise and fidelity on early broadcasting equipment and receiver sets that made listening to symphonies and operas less than pleasant. Some people thought it was because listening to a radio broadcast was just different from the way music was perceived when at a concert hall or music venue. The philosophical and aesthetic milieu surrounding what Germans then called "electrical music" became one of the intellectual cornerstones from which the studio in Cologne was created, and these minds thought that a new form of music should be created specifically for the radio medium. This idea for a new musical aesthetic came to be known as *rundfunkmusik*, or radio-music, and *neue sachlichkeit,* or the new objectivity.

In 1930, Trautwein was hired as a lecturer on the subject of electrical acoustics for the RVS. One of the other goals of the institution was to create new musical instruments that specifically catered to the needs of radio, with an overarching goal to create new tonalities that would electrify the airwaves and sing out in greater fidelity inside people's homes. It was at RVS that Trautwein collaborated with the composers Paul Hindemith, Georg Schünemann, and the musician Oskar Sala to create his instrument, the trautonium, a synthesizer that creates a vibrato by pressing down on a metal plate.

Another objective Trautwein had during his time at RVS was to analyze problems around the electronic reproduction and transmission of sound, like Harvey Fletcher and others had at Bell Labs. Unlike the people at Bell, the RVS was specifically part of a music conservatory, and though they also had the goal of clarifying speech, they were more interested in electronic music. It took Bell Labs until the 1950s to get in on that game.

One of the aims of the trautonium was to be an instrument that could be used in the home among family members for what the Germans called *hausmusik*. In addition to being small enough to fit on a table top, they wanted it to be able to mimic the sounds of many other instruments, similar to an organ. To achieve this, they worked with various resistors and capacitors and employed a glow lamp circuit to create the fundamental frequencies, which could be altered by changes in resistance and capacitance. Trautwein also added additional resonance circuits to his design that were tuned to different frequencies. He connected these to high and low pass filters that could then create formants, or concentrations of acoustic energy with a given band of resonance. All this control over the sound led to the ability to create tonalities that could be familiar and traditional, or very unusual.

With the trautonium, changes in tone color were made available with the turn of a dial. A new sound could be dialed in just as a new station could be listened to by turning the knob of a radio tuner. Tone color wasn't static either, but changed as the sound moved through time. This is the acoustical envelope of a sound, and Trautwein took this into consideration when designing his instrument.

In their search for rich tonalities, Trautwein and his colleagues stumbled across the mystery of the vowels. Preceding Homer Dudley's vocoder by eight years, the trautonium became the first electronic instrument able to reproduce the sounds of the vowels. This led Trautwein and Sala to discover the many similarities that exist between vowel sounds and the timbre of various instruments.

Trautwein compared the oscilliograms of spoken vowel formants with those played by the trautonium and found that they conformed to each other. "The trautonium is an electrical analogy of the sound creation of the human speech organs," he wrote in his 1930 paper, "Elektrische Musik," continuing, "The scientific significance lies in the physico-phsyiological impression of the synthetically generated sounds compared with the timbre of numerous musical instruments and speech sounds. This suggests that the physical processes are related in many cases." For the first iteration of the instrument, there were knobs for changing the formants and timbre, and a pedal for changing the volume. The process it used to change the tone color was an early form of subtractive synthesis that simply filtered down an already complex waveform, rather than building one up by adding sine waves together.

On June 20, 1930, a demonstration of the trautonium was given at the New Music in Berlin festival. This was to be an "Electric Concert" and one of the main attractions was the premiere of Paul Hindemith's "Seven Trio-Pieces for Three Trautoniums," written for the instrument. On one of the three instruments, Hindemith himself played the top part, with Trautwein and Oskar Sala playing the middle voice. A piano teacher named Rudolph Schmidt played the bass portion. In 1977 Oskar Sala made a recording of these short pieces using his Mixtur-Trautonium, a later iteration of the instrument. The sound is very reminiscent of the incidental music for early episodes of *Doctor Who*, and recalls soundtracks for science fiction movies in general.

Starting in 1932, a commercial version of the instrument, dubbed the Volkstrautonium, was manufactured and distributed by the German radio equipment company Telefunken, but it was expensive and difficult to learn to play, and so remained unpopular, only selling about two a year. Although discontinued in 1938, composers remained interested in the trautonium's abilities, with Hindemith, who had acted as an advisor to Trautwein, writing the *Concertina for Trautonium and Orchestra* in 1940.

For his part, Oskar Sala became a master on the instrument and would play the virtuosic compositions by Italian composer Niccolò Paganini on it. In time, he took over the further development of the trautonium and created his own variations—the Mixtur-Trautonium, the Concert-Trautonium, and the Radio-Trautonium—continuing to champion it until his death in 2002. Famously, the sound of the birds in Alfred Hitchcock's iconic 1963 thriller *The Birds* is not sourced from real birds but from the Mixtur-Trautonium played by Sala.

In 1935, the RVS was shut down by Nazi reichsminister Joseph Goebbels, but it did not disappear entirely, as its various elements were diffused into different parts of the music school. After the war, Trautwein had a hard time getting a job because he had been a card-carrying Nazi, but he did build a few more instruments, including the Amplified Harpischord in 1936 and the Electronic Bells in 1947. A modified version of the original trautonium called the Monochord, not to be confused with the stringed instrument and learning tool of the same name, was purchased by the Electronic Music Studio at the WDR in 1951.

Harald Bode and the Heinrich Hertz Institute for Research on Oscillations

Harald Bode was the next instrument maker to place his stamp upon the Electronic Music Studio at WDR, and later added a few flourishes to the work done at the Columbia-Princeton Center for Electronic Music. The son of a pipe organ player, he studied mathematics, physics, and natural philosophy at Hamburg University before inventing his first instrument, the Warbo-Formant Organ, in 1937. A completely electronic polyphonic formant organ, new sounds could be created on it by simply adjusting its half-rotary and stop knobs.

Bode next pursued his postgraduate studies at the Heinrich-Hertz-Institut für Schwingungsforschungin, or the Heinrich Hertz Institute for Research in Oscillations (HHI), located in Berlin. At the time, the HHI was focused on high frequency radio technology, telephony and telegraphy, and acoustics and mechanics for the purposes of radio, television, sound-movie technology, and architectural acoustics. Like the RVS, the HHI was also interested in developing and promoting the idea of electronic music and radio music.

It was in this phase that Bode developed his Melodium, alongside his collaborators Oskar Vierling and Fekko von Ompteda. The Melodium looked like an organ and it could play a number of different timbres, but it was only monophonic, meaning only one key or note could be played at a time. Even so, it became popular with film score composers of the era. Since it was monophonic, it presented fewer problems with tuning than his wobbly Warbo-Formant Organ. Feeling inspired by his achievement, Bode then decided that creating electronic musical instruments would be "the task of my lifetime."

His dream was put on hold when World War II broke out in 1939. Despite the dire conflict, and the spiritual sickness in his country, Bode counted himself lucky for being able to remain in the electronics industry and avoid active military duty. He still did make things for the German project, but he wasn't a foot soldier, working instead on their submarine sound and wireless communications efforts. In the aftermath of the war, he was newly married and moved from Berlin to a small village in southern Germany where he tinkered on his next invention in his home attic laboratory. The result was the first iteration of the Melochord in 1947.

Evolving upon the Melodium design, which was too idiosyncratic and complex for mass production, the Melochord had been intended to combine melody and chord capability in one instrument. Its most interesting features were the controls for shaping formants that included various filters to attenuate the sound, ring modulation for harmonics, and the ability to generate white noise and apply attack and decay envelopes. A later version featured two separate keyboards, with one being able to control the timbre of the other. The Melochord was used and promoted on the radio and in the newspapers, where it was praised for its clear and resonant tones.

When physicist Werner Meyer-Eppler got wind of the Melochord, he started to use it in his experiments at the University of Bonn. There was a lot of skill that went into playing the Melochord, and while Meyer-Eppler experimented, Bode set his sights on making a more user friendly version called the

Polychord, which became the first in a series of synthesis type organs that Bode created.

Genesis of the Studio for Electronic Music

Just as the GRM had been built around a philosophy of transforming sound, so too was the Studio for Electronic Music of the West German Radio (WDR) built around a philosophy of synthesizing sound. Werner Meyer-Eppler was the architect of the strategies to be employed in this laboratory, and the blueprint was his book, *Elektronische Klangerzeugung: Elektronische Musik und Synthetische Sprache* (*Electronic Sound Generation: Electronic Music and the Synthetic Speech*). This philosophy placed the emphasis on building up the sounds from scratch, out of oscillators and lab equipment, in contrast to the metamorphic, transformational approach purveyed by Schaeffer and Henry with musique concrète. Tape, however, remained an essential lifeblood for both studios.

Meyer-Eppler was still lecturing at Bonn's Institute for Phonetics and Communication Research while he wrote his book, in which he made an inventory of the electronic musical instruments that had so far been developed. Then Meyer-Eppler experimented with what became a basic electronic music process, composing music directly onto tape. One of the instruments Meyer-Eppler had used in his experiments was Harald Bode's Melochord, in addition to vocoders. He encouraged his students to hear the sounds from the vocoder mixed with the sounds from the Melochord as a new kind of music.

The genesis of the Studio for Electronic Music came in part from the recording of a late-night radio program about electronic music, transmitted on October 18, 1951. A meeting of minds was held in regards to the program broadcast on the Nordwestdeutscher Rundfunk. At the meeting were Meyer-Eppler and his colleagues Herbert Eimert and Robert Beyer, among others. Beyer had long been a proponent of a music oriented more towards the exploration of tone color and texture than considerations such as melody, harmony, or form. Eimert was a composer and musicologist who had published a book on atonal music back in the 1920s, while still at school at the Cologne University of Music. He had also written a twelve-tone string quartet as part of his composition examination. For these troubles, his teacher, composer and musicologist Franz Bölsche, had Eimert expelled from the class. Eimert was devout when it came to noise, twelve-tone music, and serialism, and he became a relentless advocate who organized concerts, events, and radio shows, writing numerous articles on his passion. He eventually did graduate with a doctorate in musicology in 1931, despite the attempts by Bölsche to thwart his will.

Also at the meeting was Fritz Enkel, a skilled technician who designed a framework around which a studio for electronic music could be built. The station manager, Hans Hartmann, heard a report of the meeting and gave the go ahead to establish an electronic music studio.

Creating such a studio would give national prestige to West Germany, which took great pains after the war to be seen as culturally progressive. Having a place where the latest musical developments could be explored and created by their artists was a part of showing to the world that they were moving forward. Another reason to develop the studio was to use its output for broadcasting. At the time WDR was the largest and wealthiest broadcaster in West Germany and they could use their pool of funds to create something that would have been cost prohibitive for most private individuals and companies.

Before they even got the equipment, when they felt the studio might not even get off the ground, they made a demonstration piece to broadcast and show the possibilities of what else might be achieved. Studio technician Heinz Schütz was tapped to make this happen, even though he didn't consider himself a composer or musician. The fact that a non-musician was the first to demonstrate the potential of making music in an electronic studio is apropos of the later development of the field, when people like Joe Meek and Brian Eno, who also didn't call themselves musicians, nonetheless made amazing music with the studio as their instrument.

Made with limited means, using just what they had available, the piece by Schütz was titled "Morgenröte" ("The Red of Dawn") to signify the beginning of their collective efforts. With little to work with except tape, test equipment, and recordings of Meyer-Eppler's previous experiments with the Melochord and vocoders, Eimert and Beyer "remixed" these experiments while they got their setup established. The process of working with the tapes and test equipment gave them the experience and confidence they needed for further work in their sound creation laboratory.

As the studio came together piece by piece, Eimert and Beyer put together some other sound studies, small examples of the direction they hoped to do further work in, which they premiered at the Neues Musikfest (New Music Festival) on May 26, 1953, in the broadcasting studio of the Cologne Radio Centre. The event marked the official opening of the WDR studio. Put together quickly, the pieces played did not live up to the standards Eimert had set for the studio, causing a falling out between him and Beyer, who thought they were adequate enough and resigned the next year.

Eventually Bode's Melochords and Trautwein's Monochord were acquired and modified specifically for use in the studio, which is when the studio really got cooking. In tandem with those machines they used electronic laboratory equipment such as noise and signal generators, sine wave oscillators, band pass filters, octave filters, and pulse and ring modulators. Oscilloscopes were used to look at sounds. Mixers were used to blend them together. There was a four-track tape recorder they used to synchronize and join sounds that had been recorded separately, a then-new technique developed from Meyer-Eppler's ideas. The mixer had a total of sixteen channels divided into two groups of eight. There was a remote control to operate the four- track and the attached octave filter. A cross-plug busbar panel served as a central locus where all the other inputs and outputs met. Connections could be switched with ease between instruments and sound sources, as if one were transferring a call at a telephone switchboard.

With the concept of a radio laboratory now a reality, electronic music was ready to be transmuted from the raw electrons forged within its crucible of equipment. Soon enough, a burgeoning genius named Karlheinz Stockhausen would showcase its massive potential with the genre's first classic recording.

Karlheinz Stockhausen's Studies in Electronics

Karlheinz Stockhausen was born on August 22, 1928, in a large manor house called "The Castle" by locals in the village of Mödrath, Germany. His father Simon was a school teacher, and his mother Gertrud had been born into a family of prosperous farmers. His sister Katherina was born the following year, and a brother Hermann-Josef the next. He grew up around music by way of his mother playing piano and singing around the house, but when she suffered a mental breakdown and was subsequently institutionalized in 1932, followed by the death of his brother a year later, the music stopped. In 1941, his mother was murdered in a gas chamber by the Nazi regime after being deemed socially or physically defective, or what the fascists called a "useless eater." A version of this episode was later dramatized in Stockhausen's first opera, *Donnerstag aus Licht (Thursday from Light).*

Around the age of seven, Stockhausen began the early stages of his musical training with piano lessons from the organist at the Altenberger Dom, or Abbey Church of Altenberg, where his family had moved. When his father married the family housekeeper, who gave birth to his two half-sisters, he left home for boarding school in 1942, where he continued to learn music, adding oboe and violin to his studies. In 1944, Stockhausen was forced to join the armed forces as a stretcher bearer, working for the hospital in Bedburg. During this time he played piano for the wounded on both sides. In February of 1945, he saw his father for the last time, who was sent to the Eastern Front and thought to have been killed in action in Hungary.

Between the deaths of his parents, and all the horrors and carnage he had seen during the war, Stockhausen was left with a strong aversion to war and its atrocities. His father was Nazi fanatic and liked to blast the militaristic marches and patriotic music of the fascist regime on the radio. Stockhausen hated these sounds thereafter, and felt that such strict types of rhythms had been used to goad people into complacence and compliance. Seeking solace in the rituals and music of the Catholic church, his sense of spirituality expanded to encompass other world traditions, but his native Christianity was always a touchstone, albeit one that he took to more as a mystic rather than a fundamentalist. In a similar way, he left behind the comforts of traditional music to explore the fringes of the avant-garde.

After the war, between 1947 and 1951, Stockhausen studied music at the Cologne Conservatory of Music, taking musicology, philosophy, and German classes at the University of Cologne, in addition to traveling with the stage magician Alexander Adrion as an accompanying pianist. Towards the end of this period of study, he met Herbert Eimert and Werner-Meyer Eppler.

Stockhausen had often thought of being a writer, having been touched by the work of Herman Hesse and Thomas Mann, whose novels *The Glass Bead Game* and *Dr. Faustus* both deal with music. In particular it was the mystical philosophy of music in Hesse's novel, and how it could be related to other bodies of knowledge, that became a model for the work he would go on to produce, providing a lasting influence.

In 1951, Stockhausen went to the avant-garde version of band camp, the annual Darmstadt summer course for contemporary classical music started in 1946, gathering together composers and musicians for concerts and lectures over a period of several weeks. At Darmstadt he encountered the music of Olivier Messiaen and was further inspired. He began studying and composing serial music of his own, and wrote his early pieces "Kreuzspiel" and "Formel." In January of 1952, he went to Paris to study under Messiaen, giving him the chance to meet his contemporary Pierre Boulez and see firsthand what Pierre Schaeffer was getting up to with musique concrète.

While hanging about with Boulez in Paris, he also met composers Jean Barraqué and Michel Philippot, all of whom were investing their time and efforts to create works of musique concrète at GRM. As his year in France progressed, Stockhausen was finally given permission to work in the studio, on the limited basis of recording natural sounds and percussion instruments for their tape library. In December, Stockhausen was given the go-ahead to make a piece of his own and produced "Konkrete Etude" making him the first non-French composer to use their resources.

The source sounds came from a prepared piano that were cut into fragments and spliced back together, then transposed using the phonogène. It took Stockhausen twelve days to make something the length of a pop song, at three minutes and ten seconds, though there is nothing pop about the result. The process caused him to become disenchanted with pure musique concrète, as the effort it took to make the work was substantial and the end result was not something he was satisfied with. He would continue to incorporate elements of musique concrète into many future works, but they were not solely within the domain of the musical style. This piece was only released with his approval in 1992 as part of a collection of his early electronic pieces.

As 1953 rolled around, Eimert invited Stockhausen to become his assistant in the WDR studio. Soon after his arrival in March, he determined that the Monochord and Melochord were useless when it came to his ambition to totally organize all aspects of sound, including the timbre. Only the humble sine-wave generator or beat-frequency oscillator would be able to do what he envisioned. He asked for these machines from Fritz Enkel, head of the calibration and testing department, who was beside himself. The station had spent a pretty penny—120,000 Deutschmarks—on their two showpiece instruments, and Enkel was skeptical of Stockhausen's ability to accomplish his task with just this limited kit, saying, "It will never work!" This was to become a refrain throughout Stockhausen's career of his ambitious projects. His reply stood him well for the rest of his career, "Maybe you're right, but I want to try it all the same."

The devices he used to create what became "Studie I" were all originally used for the calibration of radio equipment. Here they were put into the service of art. His idea was to build a piece totally from scratch with sine wave oscillators, following the serial organization of sounds, with added reverb to give a sense of spatialized sound.

Stockhausen's early pieces were as much an exploration of musical mathematics and acoustic science as they were novel pieces of new music made on tape with lab equipment. Behind these compositions is the work of Hermann Helmhotz, and behind him that of George Simon Ohm, and behind him Joseph Fourier, all of whom provided the intellectual additives necessary to synthesize Stockhausen's new music. "Studie I" can be heard as a musical-scientific exploration of Joseph Fourier's ideas about sine waves and how they correspond to the harmonic of a common fundamental. It can also be heard as a further exploration of Ohm's Acoustic Law, which states that a musical sound is perceived by the ear as a set of constituent pure harmonic tones. He began his musical study with a question, theorizing,

> *The wave-constitution of instrumental notes and the most diverse noises are amenable to analysis with the aid of electro-acoustic apparatus: is it then possible to reverse the process and thus to synthesize wave-forms according to analytic data? To do so one would ... have to take and combine simple waves into various forms...*

A sine tone made with electronics contains no overtones, since it is able to be made with just a single frequency. In this respect, the sine tone can be considered the prima materia, or first matter, in the radiophonic laboratory, the basic building block required to create the magnum opus. Using the tape machines, he recorded different frequency sine waves at different volumes, mixing them together to build up new synthesized timbres in a process of manual additive synthesis. "Studie I" became the first composed piece of music using this laborious additive synthesis method. The sound of it is reminiscent of the sound of a modem, back in the days of dial-up internet, only slowed down and with some of the harsher noise filtered out. Small bursts of tone color are like pressing the numbers on a phone. These are sped up and slowed down in various ways, giving the effect of bells being struck. Some ring for a longer duration, others are higher pitched and fade fast or are cut off abruptly.

Stockhausen said the piece was "the first composition with sine tones." In this respect the first piece of pure electronic music showed his devotion to the electron as a kind of musical unit unto itself. Looking at it another way, he chose this method to differentiate himself from what Schaeffer and Henry were doing with recorded sounds, what John Cage was doing with prepared pianos, and what others were doing with the proto-synthesizers. Yet he applied the tool kit of musique concrète to his effort, doing such things as running tapes backwards, speeding them up, slowing them down, and fading them in and out. The idea behind the piece was to start at the center of the human auditory range and move outwards in both directions, to the limits of perceptible pitch, with the highest and lowest pitches getting softer and shorter as the piece progresses. The work is organized around justly intoned ratios, or whole number ratios of sound frequencies. In this case the ratios are all taken from a 5:4 major third, giving "Studie I" its metallic, bell-like quality.

In "Studie II," Stockhausen explored the serial treatment of timbre. He again used sine tones, choosing a combination of five whose frequencies are all related to each other by being the twenty-fifth root of different powers of 515. This amounts to a close approximation of the Golden Ratio, making it hard to think he came to those numbers and powers just by chance. (He later used the Fibonacci sequence as a time signature in his piece, "Klavierstucke IX," and incorporated many other mathematical codes, ciphers, magic squares and allusions into his compositions, synthesizing music, math and other fields of knowledge together into a kind of glass bead game as Hesse had pointed to in his novel.) The method of combining these tones differs from "Studie I" in that here he plays them back-to-back in a reverb chamber and records the result.

As Karlheinz Stockhausen got comfortable working in the studio, the "Konkrete Etude" and "Studies" series comprised a masterful warm-up act to much bigger things to come.

Gesang der Jünglinge

There is a mystery in the sounds of vowels. There is a mystery in the sound of the human voice as it is uttered from the mouth and born into the air. And there is a mystery in the way electrons, interacting inside an oscillating circuit, can be synthesized and made to sing. Karlheinz Stockhausen set out to investigate these mysteries of human speech and circuitry as a scientist of sound, using the newly available radiophonic equipment at the WDR's Studio for Electronic Music. The end result of his research was bridged into the vessel of music, giving the ideas behind his inquiries an aesthetic and spiritual form. In doing so, he unleashed his electroacoustic masterpiece "Gesang der Jünglinge" ("Song of the Youths") into the world.

Part of Stockhausen's inspiration for "Gesang der Jünglinge" came from his studies of linguistics, phonetics, and information theory with Meyer-Eppler at Bonn between 1954 and 1956. The other part came from his spiritual inclinations. At the time of its composition, Stockhausen was a devout Catholic, and his original conception for the piece was for it to be a sacred electronic Mass, born from his personal conviction.

According to Stockhausen's official biography, he had asked Eimert, his other mentor, to write to the Diocesan office of the Archbishop for permission to have the proposed work performed in the Cologne Cathedral, the largest Gothic church in Northern Europe. The request was refused on grounds that loudspeakers had no place inside a church. No records of this request have been uncovered, so this story is now considered apocryphal. There are doubts that Eimert, who was a Protestant, ever actually brought up the subject with Johannes Overath, the man at the Archdiocese responsible for granting or denying such requests. In March of 1955, Overath had become a member of the Broadcasting Council, making it likely he was an associate of Eimert's. What we can substantiate is that Stockhausen did have ambitions to create an electronic Mass, and that he experienced frustrations and setbacks in his search for a suitable sacred venue for its performance, one that would be sanctioned by the authorities at the church.

These frustrations did not stop Stockhausen from realizing his sound-vision. The lectures given by Meyer-Eppler had seeded inspiration in his mind, and those seeds were in the form of syllables, vowels, phonemes, and fricatives. Stockhausen set to work creating music where voices merged in a sublime continuum, with synthetic tones that he built from scratch in the studio. To achieve the desired effect of mixing the human voice with electronics, he needed pure speech timbres. He decided to use the talents of Josef Protschka, a twelve-year-old boy chorister, who sang fragments derived and permutated from the "Song of the Three Youths in the Fiery Furnace," found in the third book of Daniel. In the biblical story, three youths are tossed into the furnace by King Nebuchadnezzar. They are rescued from the devouring flames by an angel who hears them singing a song of their faith. This story resonated strongly with Stockhausen at the time, who considered himself a fiery youth. Still in his twenties, he was full of energy, but under verbal fire and critical attack from the classical music establishment who lambasted his earlier works. Through song, "Gesang der Jünglinge" showed his devotion to the divine despite this persecution.

The electronic bedrock of the piece was made from generated sine tones, pulses, and filtered white noise. The recordings of the boy soprano's voice were made to mimic the electronic sounds: vowels are harmonic spectra seemingly based on sine tones, fricatives, and sibilants are like filtered white noise, and plosives resemble the pulses. Each part of the score was composed along a scale that ran from discrete events to statistically structured, massed "complexes" of sound. Now over sixty years old, Stockhausen's mixture of synthetic and organic textures still sound fresh, effusive of something new, even angelic. The voices are heavenly and aetheric and the electronics transform the space of the listener from the everyday to the sacred.

Stockhausen eventually triumphed over his persecution when he won the prestigious Polar Music Prize (considered by some the "Nobel Prize of music") in 2001. At the ceremony, he controlled the sound projection of "Gesang der Jünglinge" through the four loudspeakers surrounding the audience.

If Stockhausen were still alive it seems he would have been heartened to know about Ambient Church, a series of events organized by music producer Brian Sweeney starting in 2016. His Ambient Church promotes "group immersions into modern contemplative, otherworldly, and universal music through site-specific audio and visual performance." All the events are held in churches because Sweeney's aim was to bring a connection to the sacred through music, where the play of light, sound, and incense all lend themselves to the creation of a liminal space. No dogma is preached at the Ambient Church, as belief or its absence is left up to the individual. These creedless events do however fill a need for connection to divinity, however it may be conceived by each person in attendance. Sweeny has said, "music is spiritual, and if you come with an intention of finding transcendence, you'll experience it... churches were built for transcendence." Electronic musicians such as Robert Rich, Windy & Carl, and Suzanne Ciani have all played at Ambient Church events in the past. Stockhausen's vision of an electronic mass intimated this phenomena where ambient and drone electronica merges with the sacred geometry of illuminated cathedrals, bringing them to life.

Making Telemusik at NHK

Following the success of the Studio for Electronic Music in Germany, musicians in other countries started to take note. In May of 1952, Japanese composer Toshiro Mayuzumi had his mind blown at a musique concrète performance at Salle de l'Ancien Conservatoire in Paris, later commenting that, "the concert was such a shock that it fundamentally altered my musical life." On the same trip Mayuzumi visited Schaeffer's studio, and when he returned to Japan, he began to implement the techniques for a film soundtrack. Working at the studios of Tokyo radio station JOQR, he produced his first explicitly musique concrète piece, "x, y, z for musique concrète" Symbolically, the "x" portion was made up of metallic sounds, the "y" of human, animal and water sounds, and the "z" was taken from sounds of musical instruments. When it was finished, the piece premiered over the JOQR airwaves and was well received by Japan's listening audience, leading the station to invite Mayuzumi to create more music in this vein. The end product of this next effort was "Boxing," a radio play with a script written by celebrated Japanese novelist Yukio Mishima. For the work, Mayuzumi employed over 300 types of sounds, and it became a smash hit across the island nation.

That same year, a group of Japanese technicians and program producers were sent some materials by their German colleagues at the WDR. Aptly titled "Technical In-House Communications from the NWDR, 1954; Special Issue about Electronic Music," the paper explored some of the gear, techniques and musical theory being employed in Cologne.

Enter Makoto Moroi, a prolific composer who studied everything from Gregorian chants, to Renaissance and Baroque music, to twelve-tone composition and serialism. Alongside his love of traditional Japanese instruments, Moroi had a growing interest in what could be done musically with electronics. For him, music was the ocean he swam in, and many different rivers contributed to his flow. This led him on a 1955 pilgrimage to Cologne, to hang out with Stockhausen and take in the state of the art at the WDR studio over three weeks.

In the fall of 1955, the NHK followed in the footsteps of WDR and began to set up their own experimental studio in Tokyo. They acquired their own Monochord and Melochord, alongside a collection of oscillators, bandpass filters, tape machines, and the other gear that enabled Japan to start charting their own course in the world of avant-garde and electronic music.

Mayuzumi was quick to get to work, producing the first completely electronic music in Japan with his trilogy, "Music for Sine Waves by Proportion of Prime Number," "Music for Modulated Waves by Proportion of Prime Number," and "Invention for Square Waves and Sawtooth Waves," investigations that were directly influenced by Stockhausen's "Studie I" and "Studie II." A year later, in 1956, the laboratory in NHK had distilled its second piece of pure electronic music, "Variations on the Numerical Principle of 7," by Mayuzumi and Moroi. For this piece, the conceptual template of "Studie II" was acutely copied, though with a different numerical basis, this time based on a scale of 49/7, divided into forty-nine tones up to the seventh overtones. These songs all have metallic quality, and feature fast and slow tones that sound very much like bells and gongs treated with reverb and other effects. The word "clang" comes to mind, as do visions of the many sacred bells used in Shinto and Buddhist shrines all across Japan.

After these initial inquiries following the lead of their European counterparts, things started to move off in directions more thoroughly Japanese. Based on a traditional Noh play from the Muromachi period (fifteenth century), Mayuzumi created the thirty minute "Aoi-no-Ue," combining the classical style of Noh singing with electronics in place of the normal instruments and drums, creating a unique twentieth century version of the material. Ethereal warblings and metallic sine waves mix with the emotionally dramatic singers, some of whose voices also seem to have been subjected to electronic treatment at certain points across the piece's forty-five minutes.

In 1959, Mayuzumi started to explore the sonorities of traditional Japanese bells in his compositions. This resulted in a series of pieces with "Campanology," the art of bell ringing, in the title. He started this work by recording the sounds of the huge bells found at Buddhist temples all over Japan. He acoustically analyzed the sound of these bells and then made his first "Campanology," a ten-minute piece synthesized from the data retrieved from his recordings. In his "Nirvana Symphony" he called the first, third, and fifth movements by this name. Later, in 1967, when the NHK equipped an eighty-eight-string piano with magnets and pickups that could be electronically modulated, he wrote the first piece for it, "Campanology for Multipiano." Again the sounds are metallic, perhaps fitting for bodies of work created with machines.

Throughout the 1950s and into the next decade, the NHK continued to produce a variety of works by a number of composers. From its inception, Wataru Uenami, the chief of the studio, had wanted to invite Karlheinz Stockhausen over and commission him to create a work for their airwaves. He finally succeeded in January of 1966, four years after Stockhausen had taken over for Herbert Eimert as director of the WDR studio.

When he arrived in Japan, Stockhausen was severely jet lagged and disoriented, unable to sleep for several days. That's when the strange hallucinatory visions set in. Laying awake in bed one night, his mind was flooded with ideas of "technical processes, formal relationships, pictures of the notation, of human relationships, etc.—all at once and in a network too tangled up to be unraveled into one process." These musings of the night took on a life of their own, and from them he created "Telemusik."

Of Stockhausen's many ambitions, one of them was to make a unified music for the whole planet. He was able to do that in this piece, though the results sounded nothing like the "world music" or "world beat" genre often found playing in coffee houses and gift shops today. In the twenty minutes of the piece, he mixed in found sounds, folk songs, and ritual music from all over the globe, including Hungary, Spain, China, Japan, the Amazon, Sahara, Bali, and Vietnam. He also used new electronic sounds and traditional Japanese instruments to create what he called "a higher unity...a universality of past, present, and future, of different places and spaces: TELE-MUSIK." This practice of taking and combining sound sources from all over is now widely practiced across all genres of music via the art form of sampling. But for Stockhausen it wasn't simply making audio collage or taking one sample to build a song around it. Even though he used samples from existing recordings to make something different, he also developed a new audio process that he termed "intermodulation."

In his own words, Stockhausen speaks of the difference between collage and intermodulation:

> *I didn't want a collage, I wanted to find out if I could influence the traits of an existing kind of music, a piece of characteristic music using the traits of other music. Then I found a new modulation technique, with which I could modulate the melody curve of a singing priest with electronic timbres, for example. In any case, the abstract sound material must dominate, otherwise the result is really mishmash, and the music becomes arbitrary. I don't like that... [I used] the chant of monks in a Japanese temple with Shipibo music from the Amazon, and then further imposing a rhythm of Hungarian music on the melody of the monks. In this way, symbiotic things can be generated, which have never before been heard.*

Stockhausen kept the pitch range of "Telemusik" deliberately high, between six and twelve kilohertz, so that the intermodulation could project sounds downwards occasionally. He wanted some of the sections to seem "far away because the ear cannot analyze it" and then abruptly it would enter "the normal audible range and suddenly became understandable." The title of the piece comes from Greek *tele*, meaning afar or far off. As in with the telephone or the television, the music works consistently to bring what was "distant" close up. Cultures that were once far away from each other can now be seen ear-to-ear and face-to-face, brought together by the power of telecommunications systems, new media formats, and new music. By using recordings of traditional folk and ritual music from around the world, Stockhausen brought the past into the future and mixed it with electronics.

To accomplish all this at the NHK studio, he used a six-track tape machine and a number of signal processors including high and low-pass filters, amplitude modulators, and other existing equipment. Stockhausen also designed a few new circuits for use in the composition, one of which was the Gagaku Circuit, named after the Japanese Gagaku orchestra music it was designed to modulate.

On April 25, 1966, the first public performance of "Telemusik" took place at the NHK studios in Tokyo. Stockhausen dedicated the score to the spirit of the Japanese people. After his visit, the experimental music germ continued to spread, and the composers who were already in on the game challenged themselves with bolder, more technical and ambitious pieces.

Hymnen

"Telemusik" prepared Stockhausen for his next monumental undertaking, "Hymnen" ("Anthems"), made at the WDR studio. Started before "Telemusik" but set aside while he was in Japan, "Hymnen" is a mesmerizing elaboration of the studio technique of intermodulation first mastered at NHK. It is also a continuation of his quest to make a form of world music at a time when the people around the planet were becoming increasingly connected in Marshall McLuhan's so-called "global village." To achieve this goal, he incorporated forty national anthems from around the globe into one composition. To start, he collected 137 national anthems by writing to radio stations in those countries and asking them to send him their recordings.

The piece has four sections, though it was first slated for six. These anthems from around the world are intermodulated into an intricate web of sound lasting around two hours long. Thrown into the kaleidoscopic mix are all manner of other sounds, produced from the entire toolkit of the WDR studio, alongside added sounds from shortwave radio. These radio sounds make the entire recording sound as if you are tuning across the bands of a world receiver radio, hearing the anthems of different countries as interval signals, colliding with each other and causing transformations as the two signals meet. In the audio spectrum, and in the radio spectrum, borders and boundaries are porous, permeable.

The point of all this was, in Stockhausen words,

> *to imagine the conception of modulating an African style with a Japanese style, in the process of which the styles would not be*

> *eliminated in order to arrive at a supra-style or a uniform international style - which, in my opinion, would be absurd. Rather, during this process, the original, the unique, would actually be strengthened and in addition, transformations of the one into the other, and above all two given factors in relation to a third would be composed. The point is to find compositional processes of confrontations and mixtures of style - of intermodulations - in which styles are not simply mixed together into a hodge podge, but rather in which different characters modulate each other and through this elevate each other and sharpen their originality.*

As with "Telemusik," his aim was to go beyond what he thought of as mere collage, or what in the early 2000s might have been called a mashup. The combination of the different materials is only the first step. When each of the elements interacts with another, it ends up being transformed, changed by the association, distilling something new from the alembic of creativity.

Just as "Hymnen" mixes different anthems together, it also fuses musique concrète with electronic music. "Hymnen" can be heard as just this recorded tape piece, but he also wrote a symphony version where the tape is played by a sound projector, or diffusionist, with a score for the accompanying orchestra. This shows his tenacity in using all manner of music making tools, and intermodulating these with one another.

Hymnen ends with a new anthem for a utopian realm called "Hymunion," a mixture of the words "hymn" and "union." And isn't it nice to imagine that worldwide hymunion can be reached through the shared communion that comes from truly listening to each other.

Artikulations

Gyorgi Ligeti, the Hungarian-Austrian composer famous for the use of his compositions in the soundtrack to Stanley Kubrick's *2001:A Space Odyssey*, was one of the many composers who spent some time at the WDR studio to create electronic works. Ligeti and his wife had recently fled Hungary for Vienna in 1956, after the uprising against the People's Republic that had been quickly quashed by the Soviets. Then they made their way to Cologne, where he met Karlheinz Stockhausen and Gottfried Michael Koenig. In the summer, he attended the Darmstadt courses and started working in the studio.

Ligeti, like the others in the Cologne milieu, came under the influence of Werner Meyer-Eppler's ideas and decided to write a work that would address "the age-old question of the relationship between music and speech." The piece was composed to be an imaginary conversation of ongoing monologues, dialogues, voices in arguments, and chatter.

He first chose different types of noise to create artificial phonemes, the distinct sound units of speech, by making recordings and grouping them into a number of categories. Then he made a formula to determine the tape-length of each type. After this, he used aleatoric methods to randomly combine them into what would become the sonic articulation of words. The work was realized in 1958 with the help of Cornelius Cardew, himself an assistant of Stockhausen. In it, Ligeti created a kind of artificial polyglot language full of strange whispers, enunciations, and utterances. "Artikulation" is abstract and cybernetic, with segments that sound like rapid bursts of conversation, and others that sound like the slow speech of someone who gives every word due thought, all washed in a bath of melodious reverb.

"Artikulation" was just one of many notable works produced at WDR. Composer Gottfried Michael Koenig was one of the studio technicians who created many key pieces there, such as "Klangfiguren II" (1955), "Essay" (1957), and "Terminus I" (1962). Naim June Paik moved from Korea to Cologne in 1958 to work at the studio, where he became interested in the use of televisions as a medium for making art, going on to become a pioneer of video art. Others who made use of the studio included English experimental composer Cornelius Cardew and Holger Czukay, co-founder of the krautrock band Can.

As the 1960s rolled into the 1970s, new electronic music equipment became available and the WDR studio received a bit of an overhaul under Stockhausen's direction. It was in this era that they obtained an EMS Synthi 100 as part of their laboratory setup, changing the game once again.

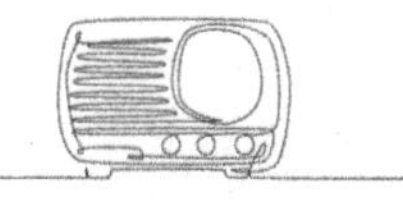

9: SONIC CONTOURS: THE COLUMBIA-PRINCETON ELECTRONIC MUSIC CENTER

Otto Luening and Vladimir Ussachevsky

In America, the laboratories for electronic sound took a different path of development, first emerging out of the universities and the private research facility of Bell Labs, rather than the public broadcasters of Europe and Japan. Taking the lead was a group of composers at Columbia and Princeton who banded together to build the Columbia-Princeton Electronic Music Center (CPEMC), the oldest dedicated institution for making electronic music in the United States. Otto Luening, Vladimir Ussachevsky, Milton Babbitt, and Roger Sessions all had their fingers on the switches in creating the studio.

Otto Luening was born in 1900, in Milwaukee, Wisconsin, to parents who had emigrated from Germany. His father was a conductor and composer, and his mother a singer, though not in a professional capacity. After his family moved back to Europe when he was twelve, Luening ended up studying music in Munich. At age seventeen he went to Switzerland, where at the Zurich

Conservatory he came into contact with Italian Futurist composer Ferruccio Busoni. Busoni was himself a devotee of Bernard Ziehn and his "enharmonic law," which stated that "every chord tone may become the fundamental." Luening picked this up and put it under his belt.

Luening eventually went back to America and worked at a slew of different colleges, and began to advocate on behalf of the American avant-garde. This led him to assist Henry Cowell with the publication of the quarterly *New Music*, while taking over their in-house label, New Music Quarterly Recordings, which put out seminal records from those inside the new music scene. It was 1949 when Luening went to Columbia for a position in the philosophy department and met Vladimir Ussachevsky.

Ussachevsky had been born in Manchuria in 1911, to Russian parents. In his early years he was exposed to the music of the Russian Orthodox Church and a variety of piano music styles, as well as the sounds from the land where he was born. He gravitated to the piano, gaining experience as a player in restaurants and as a live improviser soundtracking silent films. In 1930 he emigrated to the United States, went to various schools, served in the army during World War II, and eventually ended up under the wing of Otto Luening as a postdoctoral student at Columbia University, where he in turn became a professor.

In 1951, Ussachevsky convinced the music department to buy a professional Ampex tape recorder. When it arrived it sat in its box for a time, as he was apprehensive about opening it up and putting it to use. "A tape-recorder was, after all, a device to reproduce music, and not to assist in creating it," he later said of the experience. When he finally did start to play with the tape recorder, the experiments began. He figured out what it was capable of doing, first using it to transpose piano pitches.

One day Ussachevsky got to talking with Peter Mauzey, an electrical engineering student who worked at the university radio station, WKCR. Mauzey was able to give some technical pointers for using the tape recorder, in particular showing him how to create feedback by making a tape loop that ran over two playback heads. The possibilities inherent in tape opened up a door for Ussachevsky, and he became enamored with the medium, well before he'd ever heard of what Pierre Schaeffer and his crew were doing in France, or what Stockhausen and company were doing in Germany.

Some of these first pieces that Ussachevsky created were presented at a Composers Forum concert in Columbia's McMillan Theater on May 9, 1952. The following summer Ussachevsky was joined by Luening in presenting some of his tape music at another composers conference in Bennington, Vermont. Luening was a flute player, and they used tape to transpose his playing into pitches that were physically impossible for an unaided human, adding further effects such as echo and reverb.

After these demonstrations, Luening got busy working with the tape machine himself and started composing a series of new works at Henry Cowell's cottage in Woodstock, New York, where he had brought up the tape recorders, microphones, and a couple of Mauzey's devices. These compositions included his "Fantasy in Space," "Low Speed," and "Invention in Twelve Tones," in addition to Luening recording parts for Ussachevsky to use in his tape composition, "Sonic Contours." In these pieces, echo and delay effects created with tape loops are explored in playful symbiosis with flutes and piano. "Low Speed" shows the fun that can be had in shifting sounds down. These recordings differ from what was coming out of GRM in the way they don't include found-sounds made from objects or the environment, and from what was being done at the WDR studio by the absence of sine waves and other music made on test equipment.

In November of 1952, esteemed British conductor Leopold Stokowski premiered these pieces, along with ones by Ussachevsky, in a concert at the Museum of Modern Art, placing them squarely in the experimental tradition and helping the tape techniques to be seen as a new medium for music composition. Thereafter, the library of rudimentary equipment that began at the CPEMC would grow and be moved around from place to place. Sometimes it was in New York City, at other times Bennington, or at the MacDowell Colony in New Hampshire. There was no specific space and home for their gear.

The Louisville Orchestra in Kentucky wanted to get in on the new music game and commissioned Luening to write a piece for them to play. He agreed and brought Ussachevsky along to collaborate with him on the work, which became the first composition for tape-recorder and orchestra. To fully realize it, they needed additional equipment: two more tape-recorders and a filter, none of which were cheap in the 1950s, so they secured funding through the Rockefeller Foundation. After their work was done in Louisville, all of the tech they had so far acquired was assembled in Ussachevsky's apartment where it remained for three years. It was at this time, in 1955, they sought a permanent home for the studio through the help of Grayson Kirk, president of Columbia. He was able to help put them in a small two-story house that had once been part of the Bloomingdale Asylum for the Insane and was slated for demolition.

Here they produced the compositions "Metamorphoses" and "Piece for Tape Recorder," along with works for an Orson Welles production of *King Lear*. These efforts paid off when they garnered the enthusiasm of historian and professor Jacques Barzun who championed their work and gained them further support. With additional aid from Kirk, Luening and Ussachevsky were given a stable home for their studio inside Columbia's McMillin Theatre.

Having heard about what was going on in the studios of Paris and Germany, the pair wanted to check them out in person, see what they could learn, and possibly put to use in their own fledgling studio. Once again they were able to do this on the Rockefeller Foundation's dime. When they came back, they would soon be introduced to a machine that, in its second iteration, would go by the name ofVictor.

The Microphonics of Harry F. Olson

One of Victor's fathers was a man named Harry Olson. A native of Iowa, born in 1900, with a knack for electronics and all things technical at an early age, Olson was encouraged by his parents, who provided the materials necessary to build a small shop and lab where he tinkered with model airplanes and homebrewed steam engines. Olson got interested in ham radio, constructed his own station, demonstrated his skill in Morse Code, and obtained his amateur license. All of this curiosity, hands-on experience, and diligence served him well when he went on to pick up a bachelors in electrical engineering, followed by a master's degree with a thesis on acoustic wave filters, and topped off with a PhD in physics, all from his home state University of Iowa.

While working on his degrees, Olson came under the tutelage of Dean Carl E. Seashore, a psychologist who specialized in the fields of speech and stuttering, audiology, music, and aesthetics. Seashore was interested in how different people perceived the various dimensions of music and how ability differed between students. In 1919 he developed the Seashore Test of Music Ability, which set out to measure how well a person could discriminate between timbre, rhythm, tempo, loudness, and pitch. A related interest was in how people judged visual artwork, leading him to work with Dr. Norma Charles Meier to develop another test on art judgment. All of this work helped Seashore eventually receive financial backing from Bell Laboratories.

Another one of Olson's mentors was the head of the physics department, G.W. Stewart, under whom he did his work on acoustic wave filters. Between Seashore and Stewart's influence, Olson developed a keen interest in the areas of acoustics, sound reproduction, and music. With his advanced degree, and long history of experimentation in tow, Olson headed to the Radio Corporation of America (RCA), where he became a part of the research department in 1928. After putting in some years in various capacities, he was put in charge of the Acoustical Research Laboratory in 1934. Eight years later, in 1942, the lab was moved from Camden to Princeton, New Jersey, with facilities that included an anechoic chamber that was, at the time, the largest in the world, along with a reverberation chamber and ideal listening room. It was in these settings that Olson went on to develop a number of different types and styles of microphones. He developed microphones for use in radio broadcast, for motion picture use, directional microphones, and noise-canceling microphones. Alongside the mics, he created new designs for loudspeakers.

During World War II, Olson was put to work on a number of military projects. He specialized in the area of underwater sound and anti-submarine warfare, but after the war he got back to his main focus of sound reproduction. Taking a cue from Seashore, he set out to determine a listener's preferred bandwidth of sound when sound had been recorded and reproduced. To figure this out he designed an experiment where he put a live orchestra behind a screen fitted with a low-pass acoustic filter that cut off the high-frequency range above 5,000 hertz. This filter could be opened or closed, the bandwidth full or restricted. Audiences who listened, not knowing when the concealed filter was opened or closed, had a much stronger leaning towards the open, full bandwidth listening experience. They did not like the sound when the filter was activated. For the next phase of his experiment, Olson switched out the orchestra, whom the audience couldn't see anyway, with a sound-reproduction system with loudspeakers located in the position of the orchestra. They still preferred the full-bandwidth sound, but only when it was free of distortion. When small amounts of non-linear distortion were introduced, they preferred the restricted bandwidth. These efforts showed the amount of extreme care that needed to go into developing high-fidelity audio systems.

In the 1950s, Olson stayed extremely busy working on many projects for RCA. One included the development of magnetic tape capable of recording and transmitting color television for broadcast and playback. This led to a collaboration between RCA and the 3M company, reaching success in their aim in 1956.

The RCA Mark I Synthesizer

Claude Shannon's 1948 paper, "A Mathematical Theory of Communications," put the idea of information theory into the heads of everyone involved in the business of telephone and radio. RCA had put large sums of money into their recorded and broadcast music, and the company was quick to grasp the importance and implications of Shannon's work. In his own work at the company, Olson was a frequent collaborator with fellow senior engineer Herbert E. Belar, working together on theoretical papers and practical projects. On May 11, 1950, they issued their first internal research report on information theory, "Preliminary Investigation of Modern Communication Theories Applied to Records and Music." Their idea was to consider music as math. This in itself was not new, and can indeed be traced back to the Pythagorean tradition of music. To this ancient pedigree, they added the contemporary twist in correlating music mathematically as information. They realized that, with the right tools, they could generate music from math itself, instead of from traditional instruments. On February 26, 1952, they demonstrated their first experiment towards this goal to David Sarnoff, head of RCA, and others in the upper echelons of the company by making the machine they built perform the songs "Home Sweet Home" and "Blue Skies."

The officials gave them the green light to keep researching, resulting in the development of the RCA Mark I Synthesizer. The RCA Mark I was part computer, with simple programmable controls, yet the part of it that generated sound was completely analog. The Mark I had a large array of twelve oscillator circuits, one for each of the basic twelve tones of the musical scale. These were able to be modified by the synths and other circuits to create an astonishing variety of timbres and sounds.

The RCA Mark I was not a machine that could make music automatically. It had to be completely programmed by a composer. The flexibility of the machine and the range of possibilities gave composers a new kind of freedom, a new kind of autocracy, or total compositional control. This had long been the dream of those who had been bent towards serialism. The programming aspect of the RCA Mark I harkened back to the player pianos that had first appeared in the nineteenth century, even using a roll of punched tape to instruct the machine what to do. Olson and Belar had been meticulous in all of the aspects that could be programmed with their creation, including pitch, timbre, amplitude, envelope, vibrato, and portamento. It even featured controls for frequency filtering and reverb. All of this could be output to two channels and played on loudspeakers, or sent to a disc lathe where the resulting music was cut straight to wax.

The Mark I was introduced to the public by Sarnoff on January 31, 1955. The timing was great as far as Ussachevsky and Luening were concerned, as they first heard about it after they had returned from their trip to Europe. The experience had them eager to establish their own studio to work electronic music their own way. When they met Pierre Schaeffer, he had been eager to impose his own aesthetic values on the pair, and when they met Stockhausen, he remained secretive of his working methods and aloof about their presence. Excited about getting to work on their own, albeit exhausted from the rigors of travel, they made an appointment with the folks at RCA to have a demonstration of the Mark I Synthesizer.

The RCA Mark I far surpassed what Luening and Ussachevsky had witnessed in France, Germany, and the other countries they visited. With its twelve separate audio frequency sources, the synth was a complete and complex unit, and although programming it could be laborious, it was a different kind of labor than the kind of heavy tape manipulation they had been doing in their studio, and in the established studios of France and Germany.

The pair soon found another ally in Milton Babbitt, who was then at Princeton University. He too had a keen interest in the synth, and the three of them began to collaborate together and share time on the machine, which they had to request from

RCA. For three years, the trio made frequent trips to Sarnoff Laboratories in Princeton, where they worked on new music.

Milton Babbitt: The Musical Mathematician

Though Milton Babbitt was late to join the party started by Luening and Ussachevsky, his influence was deep. Born to a mathematician father in 1916 Philadelphia, and raised in Jackson, Mississippi, he became one of the leading proponents of total serialism. He had started playing music as a young child, first violin, then piano, and later clarinet and saxophone. As a teen he was devoted to jazz and other popular forms of music and started writing his own pop pieces. One summer, on a trip to visit family in Philadelphia, he met his uncle, a pianist studying music at the Curtis Institute of Music, who played him one of Schoenberg's piano compositions. The young man's mind was blown.

Babbitt continued to live and breathe music, but by the time he graduated high school he felt discouraged from pursuing it as his calling, thinking there would be no way to make a living as a musician or composer. He also felt torn between his love of writing popular songs and the desire to write serious music that came to him from his initial encounter with Schoenberg. He did not think the two pursuits could co-exist. Unable or unwilling to decide, he went to college specializing in math. After two years of this, his father helped convince him to do what he loved, and go to school for music.

At New York University, Babbitt became further enamored with the work of Schoenberg, who became his absolute hero, and the Second Viennese School of twelve-tone composers in general. In this time period he also got to know Edgar Varèse, who lived in a nearby apartment building. Following his degree at NYU at the age of nineteen, he started studying privately with composer Roger Sessions at Princeton. Sessions had started off as a neo-classicist, but through his friendship with Schoenberg did explore twelve-tone techniques as another tool to suit his own ends. From Sessions he learned the technique of Schenkerian analysis, a method which uses harmony, counterpoint, and tonality to find a broader, deeper understanding of a piece of music. One of the other methods Sessions used to teach his students was to have them choose a piece, and then write a new piece in a different style while using all the same structural building blocks.

Sessions got a job from Princeton to form a graduate program in music, and it was through his teacher that Babbitt eventually got his master's from the institution and joined the faculty in 1938. During the war years he got pressed into service as a mathematician doing classified work, dividing his time between Washington D.C. and Princeton, teaching math to those who would need it in the war effort, such as radar technicians. During this time he took a break from composing, but music never left his mind, doing musical thought experiments with a focus on aspects of rhythm and thoroughly internalizing Schoenberg's system. After the war was over, he went back to Jackson and wrote a systematic study of the Schoenberg system, "The Function of Set Structure in the Twelve Tone System." He submitted the completed work to Princeton as his doctoral thesis. Princeton didn't give out doctorates in music, only in musicology, and his complex thesis wasn't accepted until eight years after his retirement from the school in 1992.

Babbitt's thesis and his other extensive writings on music theory expanded upon Schoenberg's methods and formalized the twelve-tone, "dodecaphonic," system. The basic serialist approach was to take the twelve notes of the western scale and put them into an order called a series, hence the name of the style, also known as a tone row. Babbitt saw that the series could be used to order not only the pitch, but dynamics, timbre, duration, and other elements. This led him to pioneer "total serialism," which was later taken up in Europe by the likes of Pierre Boulez and Olivier Messiaen.

Babbitt treated music as a field for specialist research and wasn't very concerned with what the average listener thought of his compositions. This had its pluses and minuses. On the plus side, it allowed him to explore his mathematical and musical creativity in an open-ended way and see where it took him, without worrying about having to please an audience. On the minus side, not keeping his listeners in mind, and his ivory tower mindset, kept him from reaching people beyond the most serious devotees of abstract art music. This tendency was an interesting counterpoint from his years as teenager, when he was an avid writer of pop songs and played in every jazz ensemble he could. Babbitt had thought of Schoenberg's work as being "hermetically sealed music by a hermetically sealed man," following suit in his own career. In this respect, Babbitt can be considered as a true Castalian intellectual, and glass bead game player. Within the Second Viennese School there was an idea, a thread taken from both nineteenth century romanticism and adapted from the philosophy of Arthur Schopenhauer, that music provides access to spiritual truth. Influenced by this milieu, Babbitt's own music can be read and heard as connecting the players and listeners to a platonic realm of pure numbers.

Modernist art had already moved into areas that many people did not care about. And while Babbitt was under no illusion that he ever saw his work being widely celebrated or popular, as an employee of the university, he had to make the case that music was in itself a scientific discipline that could be explored with the rigors of science and made using formal mathematical structures. Performances of this kind of new music were aimed at other researchers in the field, not at a public who would not understand what they were listening to without education. Babbitt's approach rejected a common practice in favor of what would become the new common practice, where many different ways of investigating, playing, working with, and composing music can go off in different directions.

During World War II, Babbitt had met John von Neumann at the Institute for Advanced Studies. His association with Neumann caused Babbitt to realize that the time wasn't far off when humans would be using computers to assist them with their compositional work. Unlike some of the other composers who became interested in electronic music, Babbitt wasn't interested in new timbres. He thought the novelty of them was quick to wear off. Instead, he was interested in how electronic technology might enhance human capability for rhythms.

Victor

In 1957, Luening and Ussachevsky wrote up a long report for the Rockefeller Foundation, summarizing all they had learned and gathered so far as pioneers in the field. They included in the report another idea: the creation of the Columbia-Princeton Electronic Music Center. There was no place like it within the United States. In a spirit of synergy, the Mark I was given a new home at the CPEMC by RCA, making it easier for Babbitt, Luening, Ussachevsky, and the others to work with the machine. It would however soon have a younger, more capable brother, the

RCA Mark II, built with additional specifications as requested by Ussachevsky and Babbitt. They nicknamed the machine Victor, after the Victor Talking Machine Company, the record and phonograph company founded in 1901, famous for their line of Victrola players, later bought out by RCA and operated as RCA Victor.

There were a number of improvements that came with Victor. The number of oscillators had been doubled for starters, which gave it the potential to create more complex and nuanced sounds. Since tape was the main medium of the new music, it also made sense that Victor should be able to output to tape instead of the lathe discs used in Mark I. Babbitt was able to convince the engineers to fit it out with multi-track tape recording on four tracks. Victor also received noise generating capabilities, additional effect processes and range of controls, and a second tape punch input.

Outside of CPEMC, Mexican-American composer Conlon Nancarrow, who was also interested in rhythm as an aspect of his composition, had bypassed the issue of getting players up to speed with his complex and fast rhythms by writing works for player piano, punching the compositions literally on the roll. He even altered the auto-playing instrument slightly to increase its dynamic range. Nancarrow had also studied under Roger Sessions, and had met Babbitt back in the 1930s. Though Nancarrow worked mostly in isolation in Mexico City during the 1940s and 1950s, only later gaining critical recognition in the 1970s, it is almost certain that Babbitt would have at least been tangentially aware of his work composing on punched player piano rolls.

To operate Victor, a teletype keyboard was attached directly to the long wall of electronics that made up the synth. It was here the composer programmed their sonic creations by punching the tape onto a roll of perforated paper that was read by Victor and made into music. A worksheet had been devised that transposed musical notation to Victor's preferred language of binary code, controlling settings for frequency, octave, envelope, volume, and timbre in the two channels. In a sense, creating this kind of music was akin to working in encryption, or playing a glass bead game where one form of knowledge or art was connected to another via punches in a matrix grid.

With the RCA Mark II, Milton Babbitt found himself with a unique instrument capable of realizing his vision for a complex, maximalist twelve-tone music. He finally had the complete compositional control he had long sought after. For Babbitt, it wasn't so much the new timbres that could be created with the synth that interested him as much as being able to execute a score exactly in all parameters. His 1961 "Composition for Synthesizer" became a showcase piece, not only for Babbitt, but for Victor as well. This work has a rich warmth that distinguishes itself from the tinny and metallic pieces of electronic music that had previously come out of the GRM, WDR, and NHK studios. There is some fatter low-end, showing Babbitt wasn't afraid of a little bass. At just over ten and a half minutes, there are frequent moments of mirth and mischievousness that show that for all his highfalutin ivory tower antics, Babbitt didn't shy away from the humorous side of music. His masterpiece, 1964's "Philomel," saw the material realized on the synth accompanied by soprano singer Bethany Beardslee (wife of computer music pioneer Godfrey Winham). This subsequently became his most famous work, and sounds serious and a bit ear shattering with its many high frequencies. Contemporary electronic listeners would probably be happier if he'd made an instrumental mix, minus the operatics. The same year he created "Ensembles for Synthesizer," again featuring rich low end sounds, and hyperactive rhythms.

In the latter half of the sixties and into the first half of the seventies Milton Babbitt spent his time writing instrumental pieces for piano, orchestra and string quartet, some of them are paired with synthesized tape like "Correspondences" from 1967. The 1975 piece "Phonemena" is a voice and synthesizer work whose text is made up entirely of phonemes. Featuring twenty-four consonants and twelve vowel sounds, he explores a central preoccupation of electronic music: the nature of speech. As ever with Babbitt, the sounds are sung in a number of different combinations, with musical explorations focusing on pitch and dynamics. The soprano singer again takes this piece into the higher registers, and while it can be shrill, it is fascinating to listen to in combination with the wild electronic textures.

Modulation in the Key of Bode

Engineer and instrument inventor Harald Bode made contributions to CPEMC just as he had at WDR. He had come to the United States in 1954, setting up camp in Brattleboro, Vermont, where he worked in the lead development team at the Etsey Organ Corporation, eventually climbing up to the position of vice president. In 1958 he set up his own company, the Bode Electronics Corporation as a side project.

Meanwhile, Peter Mauzey had become the first director of engineering at CPEMC. Mauzey was able to customize a lot of the equipment to make it a comfortable place for composers. When he wasn't busy tweaking the systems in the studio, Mauzey taught as an adjunct professor at Columbia University, all while working as an engineer at Bell Labs. Robert Moog happened to be one of Mauzey's students while at Columbia, under whom he continued to develop his considerable electrical chops, even while never setting foot in the studio his teacher had helped build.

Bode left to join the Wurlitzer Organ Co. in Buffalo, New York, when the business hit rough waters around 1960. It was while working for Wurlitzer that Bode realized the power the new transistor chips represented for making music. This gave Bode the idea for a modular instrument with different components that could be connected together as needed. The instrument born from his idea was the Audio System Synthesiser. Using it, he could connect a number of different devices, or modules, in different ways to create or modify sounds. These included the basic electronic music components then in production: ring modulators, filters, reverb generators, and other effects. All of this could then be recorded to tape for further processing.

In 1960, Bode gave a demonstration of his instrument at the Audio Engineering Society in New York. Robert Moog was there to take in the knowledge and the scene, leading to his own work in creating the Moog.

In 1962 Bode started to collaborate with Vladimir Ussachevsky at the CPEMC, with whom he developed the Bode Ring Modulator and the Bode Frequency Shifter. These became staple effects at the CPEMC and were produced under both the Bode Sound Co. and licensed to Moog for inclusion in his modular systems. All of these effects became widely used in electronic music studios, and in popular music from those experimenting with the moog in the 1960s.

After retiring in 1974, Bode kept on tinkering on his own, creating the Bode Vocoder in 1971, which he also licensed to Moog, and his last instrument, the Bode Barberpole Phaser, in 1981.

Wired for Wireless

Milton Babbitt's works were just a few of the many distilled from the CPEMC. Not all composers were as obsessed with complete compositional control as Babbitt, and they utilized the full suite of processes and effects units available at the studio to create their works, which were plentiful. The CPEMC released more recorded electronic music out into the world than from anywhere else in North America, with just under thirty releases between 1964 and 1980, with periodic output all the way up to their latest release, according to Discogs, in 2022.

During the first few years of its operation, from 1959 to 1961, the capabilities of the studio were explored by Egyptian-American composer and ethnomusicologist Halim El-Dabh, who, if you remember, had been the first to remix recorded sounds using the effects then available to him at Middle East Radio in Cairo. He had come to the United States with his family on a Fulbright fellowship in 1948 and proceeded to study music under such composers as Ernst Krenek and Aaron Copland, before settling in Demarest, New Jersey. El-Dabh quickly became a fixture in the new music scene in New York, running in the same circles as Henry Cowell, John Cage, and Edgard Varèse.

By 1955, El-Dabh had gotten acquainted with Otto Luening and Vladimir Ussachevsky. At this point, his first composition for wire recorder was eleven years behind him, and he had kept up his experimentation in the meantime. Though he had been assimilated into the American new music milieu, he came from outside the scenes of both his adopted land and European avant-garde. As he had with "The Expression of Zaar," El-Dabh brought his love of folk music into the fold. His work at the

CPEMC showcased his unique combinations that involved extensive use of percussion, string sounds, singing, and spoken word, alongside the electronics. He also availed himself of Victor and made extensive use of the synthesizer. In 1959 alone, he produced eight works at CPEMC. These included his realization of an electronic drama titled, "Leiyla and the Poet," the perfect kind of hybrid of art music and audio drama for transmitting over the radio, featuring echoed drums reminiscent of some of Sun Ra's cosmic jazz masterpieces.

El-Dabh had said of his process that it "comes from interacting with the material. When you are open to ideas and thoughts the music will come to you." His less abstract, non-mathematical creations remain an enjoyable counterpoint to the cerebral enervations of his colleagues. He composed a few other pieces while working in the studio. These include his "Meditation in White Sound," a somnolent drone where the wind flows across an empty desert space only interrupted by the toll of distant bells. The white sound could be white noise, or white sand, that builds into a flowing storm as it unfolds. "Alcibiadis' Monologue to Socrates," sounds like a proto-industrial piece, with its scrapings of metallic objects, that sound as if they are emerging from inside a cavernous warehouse. "Electronics and the World" features insect-like chirpings and aggressive clangs as the backdrop for a spoken word recital. "Venice," is a softer piece featuring what sounds like field recordings treated by reverb, evocative of a gondola ride through the canals.

El-Dabh's experiments would go on to influence such musical luminaries as Frank Zappa, the West Coast Pop Art Experimental Band, fellow CPEMC composer Alice Shields, and sound-text poet, longtime KPFA music director, and founder of the Other Minds label and festival, Charles Amirkhanian.

In 1960, Vladimir Ussachevsky received a commission from a group of amateur radio enthusiasts, the De Forest Pioneers, to create a piece in tribute to their namesake, Lee de Forest, inventor of the electronic amplifier. In the studio Ussachevsky composed something evocative of the early days of radio and titled it "Wireless Fantasy." On an old spark generator in the W2ZL Historical Wireless Museum in Trenton, New Jersey, he recorded Morse Code signals tapped out by early radio guru Ed G. Raser. Woven into the various wireless sounds used in this piece are strains of Richard Wagner's "Parsifal"—a musical drama first played outside of Germany by Lee de Forest—treated with the studio equipment to sound as if it were a shortwave transmission.

From 1960 to 1961 Edgard Varèse utilized the studio to create a new realization of the tape parts for his masterpiece "Déserts." He was assisted in this task by Max Mathews from the nearby Bell Laboratories and the Turkish-born Bülent Arel, who came to work at CPEMC on a grant from the Rockefeller Foundation. For his part, Arel composed his "Stereo Electronic Music No. 1" and "No. 2" with the aid of the CPEMC facilities, as did his student Daria Semegen, who composed her "Electronic Composition No. 1" at the studio. Semegen's piece in particular is one of the trippiest electronic works from the era, showcasing the way the equipment can be used to create psychedelicized soundscapes, perfect for listening to late at night, eyes closed, ears open. Among the numerous other composers who went through the halls and walls of the CPEMC, some visiting, others as part of their formal education, there was Lucio Berio, Mario Davidovsky, Charles Dodge, and Wendy Carlos, just to name a few.

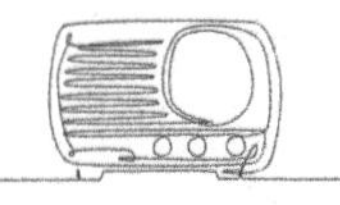

10: A NEW ATLANTIS

The Sound-Houses of Daphne Oram

As co-founder of the BBC Radiophonic Workshop—the unit created in 1958 that produced sound effects and incidental music for radio and television—Daphne Oram holds a key place in the history of electronic music. In the UK, she was one of the first proponents of musique concrète, alongside radio amateur and inveterate tinkerer F.C. Judd. Her development of the Oramics system, a sound-making technique that involves inscribing waveforms and shapes directly onto 35mm film stock, also made her an innovative inventor of a new musical technology, and the first electronic instrument designed by a woman.

Oram was born to James and Ida Oram on December 31, 1925, in Wiltshire, England. She was taught music at an early age, starting with piano and organ before moving on to composition, at which she showed considerable skill and aptitude. Her brothers introduced her to radio electronics, and they set up a system where they could transmit music over the radio around their house, constituting her first sound-house.

Daphne Oram's mother was an amateur artist. Her father was a coal merchants manager, but was also an amateur archaeologist and president of the Wiltshire Archaeological Society during the 1950s, with her childhood home a mere ten miles from the

Averbury stone circle and twenty miles from Stonehenge. It seems that her parents' interest in history and the arts lent itself to Oram's blossoming in the field of music and technology.

In 1942, at the age of seventeen, the young Oram was offered a place at the Royal College of Music, but she chose instead to take on a Junior Studio Engineer position at the BBC as a "sound balancer." She worked behind the scenes during live concerts at Albert Hall to "shadow" the musicians, ready to play a pre-recorded version of the music for broadcast in the event that the radio was disrupted by the enemy actions of the Germans—not an unlikely fear just a year after the Blitz.

Graham Wrench, the engineer who helped Oram achieve her vision for the Oramics system, had gotten to know her through his father, a professional musician with the London Symphony Orchestra, when he was still quite young. Having seen her in action at Albert Hall, he said of her work for the BBC at the time,

> *Daphne's job involved more than just setting the levels. She had a stack of records, and the printed scores of whatever pieces the orchestra was due to play. If anything went wrong in the auditorium she was expected to switch over seamlessly from the live orchestra to exactly the right part of the record!*

Another of her duties at the BBC was the creation of sound effects for radio shows. It was during this time that she started to become aware of musique concrète and the developments in electronic music. Inspired, she started to make her own experiments with tape recorders late into the night, staying to work in the studios long after her co-workers and colleagues had popped off to the pub or gone home for the evening. Cutting, splicing, playing tapes backwards, looping, speeding things up, and slowing things down were all tape techniques she learned and became expert at.

In 1948 Daphne Oram also composed an orchestral work that is now considered by some to be the first electroacoustic composition, that being a style of music making where electronics manipulate acoustic instruments. Titled "Still Point," it involved the use of turntables, a double orchestra, and five microphones. Tones shift slowly over the hour of the work creating the effect of a calm, unrippled, placid body of water, with long blocks of sustained sound that could place "Still Point" as a proto-minimalist work, if only people had heard it. Instead, the BBC rejected the piece from their programming schedule, and the composition remained unheard for seventy years. Aside from its experimental nature, there was the issue of Oram's sex. Getting a piece performed is hard enough for the unknown male composer. For women the problem was compounded by pervasive sexism, and their works were kept out of the concert halls. Many women composers turned to electronic composition as a way of bypassing the need to find ensembles willing to play a work. Nearly seventy years later, Oram's groundbreaking symphonic work was resurrected by British-Iranian composer Shiva Feshareki, who performed it with the London Contemporary Orchestra for the first time on June 24, 2016.

We Also Have Sound-Houses

Daphne Oram had already been experimenting with tape, and had even gone to the GRM studios in Paris to meet Pierre Schaeffer, when she began a crusade for the creation of a studio at the BBC dedicated to electronic music and musique concrète. Brian George was the BBC's Chief of Program Operations in 1956 when he was given a report by one of his engineers about what was going on in France with Schaeffer and Henry at the GRM, and in Germany with Meyer-Eppler, Eimert, and Stockhausen at the WDR. The report finished with a proposal for the creation of a small studio of the BBC's own, to be used for the express purpose of creating sound effects and incidental music for radio and television programs.

Oram continued her own campaign to get the studio opened, demonstrating her vision of what this music could be when she was commissioned to compose music for the play *Amphitryon 38* in 1957. This production was the first entirely electronic score for the BBC, and was made using a sine wave oscillator, self-designed filters, and a tape recorder, all in the off-hours, when the equipment wasn't otherwise being used.

The production and piece were a success and led to further commissions for Oram's electronic music. Fellow BBC colleague and electronic musician Desmond Briscoe also started to receive commissions for a number of other productions. One of the most significant was a request for electronic music to accompany Samuel Beckett's *All that Fall*, produced in 1957. The demand for electronic music was there, and the BBC finally gave in, giving Oram and Briscoe the go-ahead, and the budget, to establish the BBC Radiophonic Workshop in 1958, with Oram becoming its first director.

The focus of the Workshop was to provide sound effects and theme music for all of the corporation's output, including the science fiction serial *Quatermass and the Pit* and the radio comedy series *The Goon Show*.

One of Oram's guiding stars at the workshop came from a passage in the unfinished utopian and proto-science fiction novel, *The New Atlantis*, by Francis Bacon, first published posthumously in 1626. The novel depicts the crew of a European ship, lost at sea somewhere in the Pacific west of Peru, who eventually wash up on a mythical island called Bensalem. There isn't much plot in the book, but the setup allowed Bacon to reveal his vision of an age of religious tolerance, scientific inquiry, and technological progress. In this "New Atlantis," Salomon's House is a state-sponsored scientific institution that teases out the secrets of nature and investigates all phenomena, including music and acoustics. Bacon's book went on to form the basis for the establishment of the real life Royal Society of London for Improving Natural Knowledge. Oram found one passage in the book to be particularly prophetic, turning it into something of a mission statement for the Radiophonic Workshop. It also forms a mission statement for this book and was quoted in full in the introductory First Utterance. Here is a shortened excerpt of the passage she posted on the door of the studio:

> *We have also sound-houses, where we practice and demonstrate all sounds and their generation. We have harmonies, which you have not, of quarter-sounds and lesser slides of sounds. Divers instruments of music likewise to you unknown, some sweeter than any you have, together with bells and rings that are dainty and sweet... We represent and imitate all articulate sounds and letters, and the voices and notes of beasts and birds... We have also means to convey sounds in trunks and pipes, in strange lines and distances.*

The Disruption of a Poème Électronique

Yet even before a year was out, Daphne Oram's ambition for the sound-house she had worked so hard to establish came at loggerheads with the station executives. The inciting incident seemed to be her attendance at the *Journées Internationales de Musique*

Expérimentale exhibition at the Brussels World's Fair. It was there where she heard Edgard Varèse give a groundbreaking demonstration of his "Poème Électronique" at the Philips Pavilion.

The Philips corporation had commissioned the architect Le Corbusier to design their pavilion to showcase engineering progress, and it was Le Corbusier who came up with the title of the poem, which was to accompany a film projection of black and white photographs on themes relating to human existence selected by the architect. Somewhere between 350 and 450 loudspeakers were spread throughout the pavilion, prefiguring Beyer's later work on the Acousmonium, controlled using a number of rotary telephone dials. Out of a bank of twelve speakers, each dial could turn on five speakers at a time. With the music able to be spatialized, Varèse created a whirling kaleidoscope of sound within the cavernous acoustic structure.

Experiencing all this firsthand left Daphne Oram elated, but she soon crashed from this peak experience when she went back to work at the BBC, where she clashed with the music department who refused to put electronic music at the forefront of their agenda. When the realm of the possible smacked up against the wall of the permissible, Oram resigned from the workshop with the hope of establishing her own studio.

Oramics

Immediately after leaving the BBC in 1959, Oram began setting up her Oramics Studios for Electronic Composition in Tower Folly, inside of a former oast house (a building designed for drying hops prior to brewing) near Wrotham, Kent. The technique she created there involved the innovative use of 35mm film stock, where shapes drawn or etched onto the film strips could be read by photo-electric cells and transformed into sounds.

According to Oram, "Every nuance, every subtlety of phrasing, every tone gradation or pitch inflection must be possible just by a change in the written form." Her idea of drawing the sound to be heard was in the air at the time, following the graphic scores first done by Varèse, Stockhausen, and others.

While innovative, the Oramics technique was also expensive, both in terms of film and parts for the machine itself. Meeting the financial pressure of having her own studio, she opened it up for commercial work. Being director of the studio gave her complete control and freedom to experiment, but it also meant dealing with the stress of making works that were economically viable. She couldn't afford to be a composer like Milton Babbitt, and for the first few years she made music for commercial films, sound installations, and exhibits, as well as material for television and radio, such as the electronic sounds featured in Jack Clayton's 1961 psychological horror film, *The Innocents*. She also collaborated with opera singers and created material for concert works.

Oram's need to stay above board eased in 1962 when she was given a grant of £3,550, equivalent to £76,000 in today's money. At this point she was able to put more effort into building her drawn-sound instrument.

In 1965 Oram reconnected with Graham Wrench, a few years after she had bumped into him at the IBC recording studio where she had brought in some tape loops for a commercial. In need of an engineer and technician, and she asked Wrench if he wanted the job, so he drove down with his wife to check things out. Her studio was a mess, and while she had some skill in electronics, her gifts lay more on the creative side. She was in need of a technician who could bring her vision to life, and Wrench left his job to work for Oram.

Graham Wrench was able to help her build the system up, drawing on his experience as a radar specialist in the RAF. He started by designing a time-base for the waveform generator. To do this he needed to make his own photo-transistors, which were too expensive to buy commercially, by scraping off the paint of regular transistors.

The waveform-generator itself worked in the same fashion as an oscilloscope, but in reverse. Where an oscilloscope makes a visual representation of a sound, her Oramics system made a sound representation of a visual image inscribed on the film strip. Getting it to work was exceedingly complex, but Wrench somehow managed. The system had become very flexible in the sonics it was able to produce, with the ability to draw a sound giving an amazing freedom to creating rich envelopes of music.

Sadly Wrench, who had done so much to develop the system, was let go by Oram after she experienced an undiagnosed illness that some believed had been a brain hemorrhage. Wrench believed it was a nervous breakdown caused by her long working hours and the five hertz subharmonic frequencies caused by the Oramics machine, known to interfere with human biology. Though he later fixed the subharmonics that were emitted by adding a high pass filter, damage might have already been done. Just as the exact cause of her illness remains unknown, so too were the exact reasons she let go of the technician who manifested the system she had envisioned.

Other engineers and technicians came in and copied what Wrench had done to expand the Oramics system, while Oram continued to compose, research, and think about the implications of electronic music from a philosophical perspective. She turned her attention to the subtle nuances of sound that composers had never been able to control using traditional instruments. She applied this research to the study of perception itself, and how the human ear influences the way the brain apprehends the world. Oramics came to encompass a study of vibrational phenomena, and she divided her system into two distinct parts: the commercial and the mystical. In her detailed notebooks, Oram defined Oramics as "the study of sound and its relationship to life."

Throughout her career, Daphne Oram lectured on electronic music and studio techniques. In the early 1970s she was commissioned to write a book on electronic music. She didn't want it to become a how-to book, instead taking a philosophical, meditative approach to the subject. *An Individual Note of Music, Sound and Electronics* was published in 1972 and reissued in 2016.

Later in the '70s, Oram began a second book, *The Sound of the Past - A Resonating Speculation*, which never saw print but survives as a manuscript. In this work, the influence of her father's antiquarianism, and her own interest in the mystical side of life, can be seen. In it she speculates and muses on the subject of archaeological acoustics and proposes a theory, backed by research, suggesting that Neolithic chambered mounds and ancient sites like Stonehenge and the Great Pyramid in Egypt were used as resonators, and could be used to amplify sound. Her research suggested that ancient peoples, through their knowledge of acoustics, may have been able to use these places for long distance communication.

By the time the 1980s rolled around, Oram was engaged by the Acorn Archimedes computer company to work on the development of a software version of Oramics for their computer system, receiving a grant from the Ralph Vaughan Williams Trust. She had wished to continue the mystical side of her sound research, but the continuing financial struggles for such a project left that dream in limbo and incomplete. In the 1990s Oram suffered from two strokes that eventually led her away from her work and into a nursing home. She died in 2003.

In her book, Oram predicted the future, musing, "We will be entering a strange world where composers will be mingling with capacitors, computers will be controlling crotchets and, maybe, memory, music and magnetism will lead us towards metaphysics."

The Spherical Vortices of Delia Derbyshire

A few years after Daphne Oram stepped out of the BBC Radiophonic Workshop, another lady stepped in. Though Delia Derbyshire may not be a household name, the sound of her music is certainly embedded in the brains of several generations of science fiction fans, as it was she who realized the iconic *Doctor Who* theme song in the Workshop studios. With the original *Doctor Who* series lasting for twenty-six continuous seasons, from 1963 to 1989, the song has touched the lives of millions of people around the world.

Derbyshire had been a bright girl, learning to read and write at an early age, and training on the piano at age eight. But like many others in this book, it was radio that opened up her ears to the world, saying later, "the radio was my education." After graduating from Barr's Hill Grammar School in 1956, she was accepted by both Oxford and Cambridge, which was, in her own words, "quite something for a working class girl in the 'fifties, where only one in 10 were female." She ended up going to Girton College at Cambridge because of a mathematics scholarship she received.

Despite some success with the mathematical theory of electricity, she claimed to have not done so well in school at the time, switching her focus to include music, specializing in medieval and modern music history, while graduating with a BA in mathematics. She also received a diploma, or what the British call a licentiate, from the Royal Academy of Music in the study of pianoforte.

While in school, she developed an interest in the musical possibilities of everyday objects. This would later find its full expression in the musique concrète she would make and master at the BBC. In 1958, she also had the opportunity to visit the World's Fair in Brussels, where, just like Daphne Oram, she experienced Edgard Varèse's "Poème Électronique" installed in Le Corbusier's pavilion.

Upon finishing her schooling, Derbyshire approached the university career office for advice. The pieces had been arranged on the board of her life, but she needed help with making her next move. She told the counselor she had an interest in sound, music, and acoustics, "to which they recommended a career in either deaf aids or depth sounding." With their advice wanting, she made a move on her own and tried to get a gig at Decca Records, but was told no, as no women were employed in the recording studio of the label. Women had their set role as performers and music teachers, but the idea of women composers and sound engineers was still met with resistance in the professional music world.

In lieu of a job with Decca, she scored a position with the UN in Geneva, as a piano and math teacher to the children of various consuls and diplomats. Later she worked as an aide to Gerald G. Gross, who oversaw conferences for the International Telecommunications Union. Eventually she moved back home to Coventry where she taught at a primary school. This was followed by a brief stint in the promotions department at the music publisher Boosey & Hawkes.

The following year, in 1960, she stepped into the BBC as trainee assistant studio manager. Her first job there was working on the *Record Review*, a program where hoity-toity critics gave their highfalutin opinions on classical music recordings. Just like Daphne Oram, she had a well-developed sense of where to drop the needle on any given platter, remembering, "some people thought I had a kind of second sight. One of the music critics would say, 'I don't know where it is, but it's where the trombones come in' and I'd hold it up to the light and see the trombones and put the needle down exactly where it was. And they thought it was magic."

Of this time period, Derbyshire further elaborated, "It was very exciting, especially on the music shows. All the records had to be spun in by hand and split second timing was essential. When tapes came in I used to mark them with yellow markers to ensure that one followed another, and that there were no embarrassing gaps in between."

Not long after, she heard about the sound-house Daphne Oram had helped create, the Radiophonic Workshop, and she knew she wanted inside, developing and working in the new field of electronic and electro-acoustic music, and exploring the widest parameters of musical research. When she approached the heads of Central Programme Operation with her wish to transfer to the Workshop, they were baffled and puzzled. It wasn't a place most people sought out to work in, but rather a place people were sent or assigned, no doubt with grumbling resentment. It was a place only the eccentric, or visionary, would choose to go.

In 1962 Derbyshire got her wish and was appointed to the Radiophonic Workshop in Maida Vale. "I had done some composing but I had a running battle with the B.B.C. to let me specialise in this field," she noted in a newspaper article. "Eventually they gave me three months to prove I was good -- and I'm still here." For the next decade and a year, she gave the BBC a herculean effort in the creation of sound and music for about 200 radio and television shows, as well as films, including the voice of the ghost of Hamlet's father in a 1969 movie version of the play. Many of her pieces were used in documentaries and science fiction radio dramas.

Doctor Who

In 1963, the year after she got her wish, *Doctor Who* came to broadcast, becoming a force of nature in its own right. The theme song was one of the first on television to be made entirely with electronics, with Delia Derbyshire taking over the job of realizing the score, after being initially composed by Ron Grainer. It was a laborious process and the Radiophonic Workshop had become the perfect laboratory for the great work of sonic separation, granulation, elaboration, and final distillation of the musical substance.

To create the *Doctor Who* theme, each note was individually recorded, cut, and spliced. Some of the base materials used for the process included a single plucked string, white noise, and the harmonic waveforms of test-tone oscillators. The bass line was from the plucked string, made into a pattern by splicing it in versions that had been sped up or slowed down to create the perfect pitch, over and over again. The swoop of the lower bass layer was made through careful and calculated tweaking of the oscillator's pitch. The melody was played on a keyboard attached to a rack of oscillators while the bubbling hiss and fry of some aetheric vapor was made by filtering white noise and then arranging it in time on tape. Some of the notes were also redubbed at varying volumes to create the necessary dynamics heard in the song.

With all the basic material in the laboratory now prepared, ready with the proper pitch and volume, it all needed to be conjoined. To do this, the first step involved taking a line of music—the bass, melody, or vaporous bubbles of white noise—and trimming each note to length by cutting the tape and sticking them all together in the right order. Next, further rectifications were required, distilling these elements down further and further until a final mix was completed.

At the time, there were no multitrack tape machines to ease the process, so a method to mix it all together had to be improvised. Each separate portion of the song on individual reels of tape was played on separate tape machines with the outputs mixed together. Getting it all to synchronize was just one of the obstacles, as not all tape players play back at exactly the same speed, and not all of them stay in sync once started. A number of submixes, or distillations, were created, and these in turn synced together before the music could finally be said to be finished.

When Ron Grainer first heard Derbyshire's realization of his score, he was more than delighted and said "Did I really write this?" She replied, "Most of it." Grainer made a valiant effort to give Derbyshire credit as a co-composer of the theme, but his attempt was blocked by the bureaucrats at the BBC, who had the official policy of keeping the members of the Workshop anonymous and only giving credit to the group as a whole. Derbyshire was not credited on screen for her work until the fiftieth anniversary special of *Doctor Who*. Even so, her tenure in the Workshop was off to a grand start, and she continued to produce music for radio, television, and beyond.

Brian Hodgson, who worked with Delia Derbyshire at the Workshop, and also produced a lot of incidental music for *Doctor Who*, commented on her work on the theme:

> *It was a world without synthesizers, samplers and multi-track tape recorders; Delia, assisted by her engineer Dick Mills, had to create each sound from scratch. She used concrète sources and sine- and square-wave oscillators, tuning the results, filtering and treating, cutting so that the joins were seamless, combining sound on individual tape recorders, re-recording the results, and repeating the process, over and over again.*

Derbyshire was interviewed about the theme on a 1964 episode of the radio show, *Information Please*, and said, "the music was constructed note by note without the use of any live instrumentalists at all," and went on to demonstrate the use of various oscillators, including the workshop's famous wobbulator, which she said was "simply an oscillator which wobbles."

Due to the popularity of *Doctor Who*, her work and that of her colleagues at the Workshop for the theme and incidental music became, in many ways, the first really widespread electronic works to find a wide and enthusiastic listening audience.

Inventions for Radio

Between 1964 and1965, Delia Derbyshire got to expand her palette of sound across the canvas of radio by collaborating with playwright Barry Bermange in a series of four pieces called *Inventions for Radio*.These pieces were broadcast on the *BBC Third Programme* and involved interviews with people on the street on such heavy subjects as dreams and the existence of God, collaged against a background of electronic soundscapes and strange noises. It was a new form of documentary radio art.

Working with Bermange, the voices of the interviewees were edited in a non-linear way, creating insightful juxtapositions. For the episode on dreams she used one of her favorite musical sources, a green metal lightbulb shade being struck. The sound, as always, was later manipulated in the studio.

And even though her work for the Workshop continued to remain anonymous, her reputation as a musician and electronic composer started to spread to some of the senior officials at the communications behemoth. Martin Esslin, the BBC's Head of Radio Drama, sent a memo to Desmond Briscoe, then head of the Workshop, noting his regret that Delia Derbyshire and her co-worker John Harrison were not able to receive credit for the work they had on a production of *The Tower*. He wrote,

> *I have just been listening to the playback of the completed version of 'The Tower' and should like to express my deep appreciation for the excellent work done on this production by Delia Derbyshire and John Harrison. This play set them an extremely difficult task and they rose to the challenge with a degree of imaginative intuition and technical mastery which deserves the highest admiration and which will inevitably earn a lion's share of any success the production may eventually achieve. I only wish that it were possible for the names of contributors of this calibre to be mentioned in the credits in the Radio Times and on the air. But failing this I should like to register the fact that I regard their contribution to this production as being at least of equal importance to that of the producer himself.*

Kaleidophonics

As Delia Derbyshire's reputation grew, she began work on other projects outside the umbrella of the BBC. She joined forces with her friend and fellow Radiophonic Workshop member Brian Hodgson, along with Peter Zinovieff, the creator of the EMS synthesizer, to establish Unit Delta Plus. The purpose of this organization was to promote and create electronic music. A studio Zinovieff had built in a shed behind his townhouse at 49 Deodar Road in Putney served as their operational headquarters.

A true student of electronic music-making, Zinovieff had followed the research of Max Mathews and Jean-Claude Risset at Bell Labs, and read David Alan Luce's 1963 MIT thesis, "The Physical Correlates of Nonpercussive Musical Instrument Tones," just the kind of thing to read on a rainy day. He had both a theoretical and practical knowledge of electronic music that influenced his work.

Derbyshire, Hodgson, and Zinovieff made quite the trio, and they soon participated in a few experimental and electronic music festivals. In 1966 they demonstrated their electronic prowess at the Million Volt Light and Sound Rave. This was the same event where The Beatles had been commissioned to create an avant-garde sound piece. The Fab Four came up with the song "Carnival of Light" in response, which had its only public playing at the festival. They went into the studio to record a version of the piece in January of 1967, but it never had an official release. It prefigures the sound collage work they would later do on "Revolution 9" and features extended freak out sections, mottled drums, and many strange warblings. Psychedelic rock bands would later incorporate these kinds of avant-garde explorations into their long trippy songs, perhaps most notably by Pink Floyd on songs such as "Echoes," but picked up and explored by many other bands that would follow. "Carnival of Light" shows The Beatles' engagement with the more experimental end of music, even as it remains officially unreleased. The following year they would place Karlheinz Stockhausen on the cover of *Sgt. Pepper's Lonely Hearts Club Band*, along with their other heroes.

Though there were intervening projects for Unit Delta Plus, the next major one outside of the BBC was to mark another landmark in the history of electronic music. It all got sparked when Derbyshire and Hodgson met David Vorhaus, a classical musician trained as a bass player. He also happened to be a physics graduate and electronic engineer. Vorhaus recalls:

> *I met Brian Hodgson and Delia Derbyshire, who were then in a band called Unit Delta Plus. I was on my way to an orchestral gig when the conductor told me that there was a lecture next door on the subject of electronic music. The lecture was fantastic and we got on like a house on fire, starting the Kaliedophon studio about a week later!*

These three were another electrical storm of creative energy, together creating the Kaleidophon studio at 281-283 Camden High Street, where they made music and sound for a variety of London theaters. They also made library music, contributing many tracks to the Standard Music Library, a firm set up in collaboration with London Weekend Television (ITV) and Bucks Music Group in 1968 to provide the background and incidental music for hit TV shows. Often these recordings were made under pseudonyms, with Derbyshire's compositions credited to Li De La Russe, something of an anagram and reference to her auburn hair. A number of these songs made it onto the ITV shows *The Tomorrow People* and *Timeslip*, another sci-fi time-travel show that rivaled *Doctor Who* for viewership.

When not busy working on a commission, they were busy crafting their first album as the band White Noise. Titled *An Electric Storm*, the album is a masterpiece, spanning genres of giddy electro-pop to more austere and serious sonorities. It spans a deep emotional gamut and is an excellent and dizzying listen from start to finish. Released on Island Records, it was something of a sleeper album, or what some call a perennial seller, not selling great upon its initial release, but steadily afterwards. Considering the difficulties the band had in even getting it onto a label makes their achievement even more remarkable. Tracks like "Love Without Sound" and "Firebird" take their combined mastery of studio techniques and put them in service of the pop hook, the latter something of an electronic love ballad. "Here Come the Fleas" sounds like the novelty music championed by Dr. Demento while "The Visitation" is a nightmare soundscape that revels in a disturbing spectrality. Though the name White Noise lives on with David Vorhaus, Hodgson and Derbyshire left the project and the studio after the first album.

A number of other commissions, recordings, and events took place as the last years of the 1960s unspooled. Derbyshire made music for a film by Yoko Ono, contributed to Guy Woolfenden's electronic score for *Macbeth* produced by the Royal Shakespeare Company, and collaborated with Anthony Newley for a demo song called "Moogies Bloogies," another gem of early electropop that has never seen an official release.

In 1970 Delia worked on an episode for the TV show series *Biography* that detailed the life of Johannes Kepler, the renaissance astronomer who showed that planets orbit the sun in ellipses, not perfect circles. The episode was titled, *I Measured the Skies* and was taken from Kepler's epitaph which reads:

> *I measured the skies, now the shadows I measure, Sky-bound was the mind, earth-bound the body rests.*

In his book *Harmonices Mundi* from 1619 Kepler explored the relationships between musical harmony and the congruence of geometrical forms and physical phenomena. His book is most famous for relating his third law of planetary motion which suggests that the time it takes for a planet to orbit the Sun increases swiftly with the radius of its orbit. Medieval philosophers had spoken of the music of the spheres as metaphor. Kepler discovered actual physical harmonies in planetary motion, finding harmonic proportions in the differences between the maximum and minimum angular speeds of a traveling planet.

The TV program was featured in a newspaper article by Christine Edge where she explained that, "Kepler had interpreted the sounds made by the planets into scale notes, and Delia subjected them to her own gliding scale of electronic sounds." A few years later Delia revisited the music of the spheres, this time producing a piece for a segment on Kepler in Joseph Bronowski's 1973 TV series *The Ascent of Man*. Her short piece accompanies a simple computer graphic being shown on the screen.

Delia was in her own sphere and orbit, and as her velocity accelerated the people around started to notice its wobble.

I.E.E.100

In 1971 the Institute of Electrical Electrical Engineers (IEE) –not to be confused with the Institute of Electrical and Electronics Engineers (IEEE) in America– turned 100 years old. The BBC commemorated the anniversary with the Radiophonic Workshop in a concert event on May 19. Though Delia Derbyshire composed the piece "I.E.E. 100" for the program, the tape almost didn't survive.

Looking to the history of radio and electrical engineering for inspiration, she said of the project,

> *I began by interpreting the actual letters, I.E.E. one hundred, in two different ways. The first one in a morse code version using the morse for I.E.E. 100. This I found extremely dull, rhythmically, and so I decided to use the full stops in between the I and the two E's because full stop has a nice sound to it: it goes di-dah di-dah di-dah. I wanted to have, as well as a rhythmic motive, to have musical motive running throughout the whole piece and so I interpreted the letters again into musical terms. 'I' becomes B, the 'E' remains and 100 I've used in the roman form of C.*

Further elements of the piece included many touchstones of the history of telecommunications, from the development of electricity in communication, to the earliest telephone, to the Americans landing on the moon. She sampled the 1888 recording of William Ewart Gladstone congratulating Thomas Edison on inventing the phonograph. She sampled the voice of Lord Reith, the founder of the BBC, when he opened and closed down Savoy Hill where the BBC had their initial recording studios. Savoy Hill had been the home of the IEE starting in 1889 and opened its doors to make a space for the BBC in 1923. The voice of Neil Armstrong speaking as he stepped onto the surface of the moon was also used in her work. Describing his first impression of the piece, English composer Desmond Briscoe wrote, "The powerful punch of Delia's rocket take-off threatened the very fabric of the Festival Hall."

This was one of the events where Derbyshire's chronic perfectionism began to show itself, having a deleterious effect on her ability to finish work, despite being a professional who had tackled numerous large projects. She was working on the piece up to the night before the event, making edits, trying to make it live up to the rigorous standards she set for herself. Brian Hodgson was in charge of directing the program, and was aware that she might have a breakdown and do something to the tape, so he called upon one of the Workshop's engineers to secretly make a second copy of the final version of the work and to give it to him. Hodgson's intuition was quite correct, later recalling,

> *I said to Richard [the engineer] "Run another set in Room 12, don't tell Delia you're doing it, and that copy bring to me in the morning, because I have an awful feeling she was going to destroy the tape." And he did that. And she came in the next morning in tears, around 11 o'clock. And said, "I've destroyed the tape, what are we going to do?" I don't think she ever forgave me for that.*

Two years later, Derbyshire would leave the BBC, fed up. In an interview on Radio Scotland, she said, "Something serious happened around '72, '73, '74: the world went out of tune with itself and the BBC went out of tune with itself... I think, probably, when they had an accountant as director general. I didn't like the music business."

She spent a brief time working at Brian Hodgson's Electrophon Studio before quitting that too. It was hard for her to quit radio though, as it is for many who've been hooked and tried to give it up. She got a gig working as a radio operator, saying of the time,

> *Crazy, crazy, crazy! I was the best radio operator Laing Pipelines ever had! I answered a job in the paper for a French speaking radio operator. I just had to sleep - everything was out of tune, so I went to the north of Cumbria. It was twelve miles south of the border. I had a lovely house built from stones from Hadrian's Wall. I was in charge of three transmitters in a disused quarry. I did not want to get involved in a big organisation again. I'd fled the BBC and I thought - oh, Laing's... a local family firm! Then I found this huge consortium between Laing's and these two French companies.*

By 1975 Derbyshire had stopped producing music for public consumption. According to Clive Blackburn, "in private, she never stopped writing music either. She simply refused to compromise her integrity in any way. And ultimately, she couldn't cope. She just burnt herself out. An obsessive need for perfection destroyed her."

Yet in the 1990s, Delia Derbyshire started seeing the electronic music she had championed starting to come into its own. Pete Kember, a member of the psychedelic noise rock band Spacemen 3 who also goes by the stage name Sonic Boom, sought Derbyshire out and befriended her. Having amassed a collection of synthesizers and electronic music gear, Kember was juggling two new projects called Spectrum and E.A.R. (Experimental Audio Research), in both cases making the kind of music she had been at the forefront of in previous decades. While recording his 1997 Spectrum album, *Forever Alien*, which features a track titled

"Delia Derbyshire," Kember looked her up in the phone book, later remembering,

> *We spent a lot of time talking on the phone, three or four hours a night. She taught me pretty much everything I know about the physics of sound. The basis of modular synthesis I taught myself, but stuff like the harmonic series, she'd spend endless hours helping me get it into my thick head.*

Derbyshire's life had become chaotic though. The ravages of alcohol abuse were catching up with her body, and just as she started to work on public music again with Kember's E.A.R. in 2001, she died of renal failure. A short fifty-five second collaboration they had made called "Synchrondipidity Machine (Taken from an Unfinished Dream)" was released after she had departed and was dedicated to her memory. Kember credited her with "liquid paper sounds generated using fourier synthesis of sound based on photo/pixel info (B2wav - bitmap to sound programme)."

After Derbyshire died, 267 reel-to-reel tapes and a box of a thousand papers were found in her attic. These were entrusted to Mark Ayres of the BBC, and in 2007 were given on permanent loan to the University of Manchester. Almost all the tapes were digitized in 2007 by Louis Niebur and David Butler, but none of the music has been published due to copyright complications.

Delia Derbyshire's life was an unfinished dream, and it is a shame she did not stick around long enough to see the credit that was later bestowed on her for her generous contributions to electronic music.

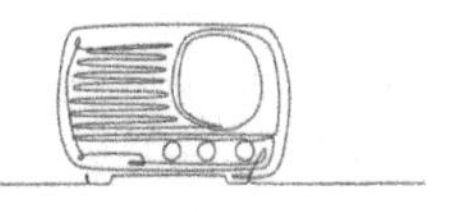

11: ECHOES OF THE BELL

Max Mathews Likes to GROOVE

In 1970, at Bell Telephone Labs, Max Mathews had started working with Richard Moore to create the GROOVE system, intended to be a "musician-friendly" computer environment. His MUSIC programs had broken incredible new ground, but the use of them leaned more towards those who were willing to learn to write code in their esoteric languages, excluding the average musician or composer of the time. GROOVE was the next step in making the computer a tool for music creators, a hybrid digital-analog system that stood for "Generating Realtime Operations On Voltage-controlled Equipment."

Bell Labs had acquired a Honeywell DDP-24 computer from MIT to use specifically for sound research. This is what GROOVE was built on. The DDP-24 was a 24-bit transistor machine that used magnetic core memory to store data and program instructions. That it had disk storage also meant it was possible for libraries of programming routines to be written, allowing the users to create customized logic patterns. A composition could be tweaked, adjusted, and mixed in real time on the knobs, controls, and keys, and in this manner a piece could be reviewed as a whole or in sections and then replayed from the stored data.

GROOVE was a first step to making computer music in real time. The setup included an analog synth with a computer and monitor. The computer's memory made it appealing to musicians who could store their manipulations of the interface for later recall, a clever workaround to the limitations of each technology. The computer was used for its ability to store the musical parameters, while the synth was used to create the timbres and texture without relying on digital programming. This setup allowed creators to play with the system and fine tune what they wanted it to do for later re-creation. The music could be adjusted and mixed in real time, giving those who worked with the system a distinct advantage over the time-consuming process needed to convert digital tapes to audio, as had been the case with computer music up to that point.

When the GROOVE system was first demonstrated in Stockholm at the 1970 conference on Music and Technology, organized by UNESCO, the event was attended by Jean-Claude Risset and Pierre Schaeffer and synthesized music by Bartok and Bach was played.

Grooving with Laurie Spiegel

One of the key composers exploring the new frontiers of science and music at Bell Labs was Laurie Spiegel. The year was 1973, and at age twenty-eight she was already an accomplished musician with a restless curiosity when she started working with the GROOVE interactive compositional software. Having honed her skill and ability through scholarly devotion to musical practice and study, she was the kind of person who could see the creative potential in the new tools the facility was creating and make something timeless.

Born in Chicago, Spiegel's musical interest had begun with stringed instruments, the ones you strum and pluck, teaching herself to play guitar, banjo, and mandolin by ear in her teens. She excelled in high school and was able to graduate early and get a jump start on a more refined education. Shimer College in Illinois had an early entrance program and she made the cut. Using Shimer as a launching board, she got into their study abroad program and left to join the scholars at Oxford University. While pursuing her degree in social sciences she decided to teach herself music notation, an essential skill if she was to start writing down her own compositions. She managed to stay on at Oxford for an additional year after her undergraduate was completed. In between classes she would commute to London for lessons with composer and guitarist John W. Durante who fleshed out her musical theory and composition. She was no slacker.

Laurie Spiegel's devotion to music continued to flourish when she came back to the states. In New York she worked briefly on documentary films in the field of social science, but the drive to compose music pushed her back onto the path of continuing education. She headed back to school again, at Juilliard, going for a master's in composition. Emmanuel Ghent, Hall Overton, and Vincent Perischetti were some of her teachers between 1969 and 1972, as was Jacob Druckman, who she ended up becoming an assistant to, and following to Brooklyn College. While there she also managed to find some time to research early American music under H. Wiley Hitchcock before completing her MA in 1975.

Besides playing guitar, Spiegel had been working with the analog modular instruments made by Buchla, Electrocomp, Moog, and Ionic/Putney. After a few years of experimentation, she outgrew these synths and started seeking something that had the control of logic and a larger capacity for memory. This exposed Spiegel to the work being done with computers and music at Bell Labs.

In 1973, while still working on her master's, she started experimenting at Bell Labs, using the system developed by Max Mathews and Richard Moore. GROOVE was to prove the perfect foil for expressing Spiegel's creative ideas. While Max Mathews was bouncing around between a dozen different departments, Spiegel was getting her GROOVE on at Murray Hill.

GROOVE took the wait off of composers who wanted to make computer music. Those who were working with software such as Mathew's MUSIC V had to do all their programming ahead of time, leave the computer to run for long hours, often overnight or even over a weekend depending on how large the job was, and come back for it to get the audio, which could be very short, as little as thirty seconds of a song. Spiegel had fallen in love with being able to hear in real time the work she was doing when she played the Buchla and Moog synths, yet she still desired more control than those devices could offer. GROOVE gave her the control *and* the interactivity she wanted, without the wait. With the nuances of the system taking up less of her time, she was able to devote more of it to making music as beautiful as the harmony of the spheres.

Harmonices Mundi

Spiegel's most famous work is also the one most likely to be heard by space aliens. A realization of astronomer Johannes Kepler's 1619 book *Harmonices Mundi* using the GROOVE system, it was the first track featured on the golden phonograph records famously placed aboard the two Voyager spacecrafts launched in 1977.

Forming a kind of time capsule, a message in a bottle sent off into interstellar space, the records contain sounds and images intended to portray the vast diversity of life and culture on planet Earth.

Carl Sagan chaired the committee that determined what contents should be put on the record, saying, "The spacecraft will be encountered and the record played only if there are advanced space-faring civilizations in interstellar space, but the launching of this 'bottle' into the cosmic 'ocean' says something very hopeful about life on this planet."

A message in a bottle isn't the most efficient way of communicating if your purpose is to reach a specific person in a short amount of time. However, if you trust in fate, providence, or the natural waves of the ocean to guide the message to whomever it is meant to be received by, it can be oracular.

Like many musicians before her, Laurie Spiegel had been fascinated by the Pythagorean dream of a music of the spheres. When she set about turning Kepler's speculative seventeenth century composition into music, she had no idea her music would actually be traveling through the spheres. Putting forth Kepler's third law of planetary motion, *Harmonices Mundi* was based on the varying speeds of orbit of the planets around the sun. He wanted to be able to hear "the celestial music that only God could hear," as Spiegel said, continuing, "Kepler had written down his instructions but it had not been possible to actually turn it into sound at that time. But now we had the technology. So I programmed the astronomical data into the computer, told it how to play it, and it just ran."

The resulting sounds aren't the kind of thing you'd typically put on your turntable after getting home after a hectic work day to relax. Yet if you listen to the piece as the product of a mathematical and philosophical exercise, and if you are a fan of drones and alternate tunings, it is enjoyable. "Kepler's Harmony of the Worlds" is a long multilayered drone, with ascending and descending pitches sliding at different speeds. These can be visualized as the planets, each in their orbit, each at their own elliptical tilt, spinning like a set of locked groove records beneath the empyrean of the distant stars. At a suitable volume the speakers will shake with the power of the low-end vibrations, the slower moving planets, conjuring up images of the massive power and size of these heavenly bodies.

Other sounds that can be heard on the Voyager Golden Records include spoken greetings from Earth-people in fifty-five languages, Chuck Berry's "Johnny B Goode," Louis Armstrong's "Melancholy Blues," and music from all around the world, from folk to classical. Each record is encased in a protective aluminum jacket, and includes a cartridge and a needle for the aliens to play it. Symbolic instructions, kind of like those for building a piece of furniture from Ikea, show the origin of the spacecraft and indicate how the record is to be played. In addition to the music and sounds, there are 115 images encoded in analog form.

Spiegel was hanging out with some friends when she received a phone call requesting the use of her music for the record.

> *I was sitting with some friends in Woodstock when a telephone call was forwarded to me from someone who claimed to be from NASA, and who wanted to use a piece of my music to contact extraterrestrial life. I said, "C'mon, if you're for real you better send the request to me through the mail on official NASA letterhead!"*

It turned out to be the real deal and not just a prank.

In 2012, Voyager I left our solar system and entered interstellar space, and it's still out there, sending back information. Spiegel says, "It's extremely heartening to think that our species, with all its faults, is capable of that level of technical operation. We're talking Apple II level technology, but nobody's had to go out there and reboot them once!"

An Expanding Universe

Laurie Spiegel explored many other ideas within the structure of the highly adaptable GROOVE system, taking naps in the Bell Labs anechoic chamber when she needed a rest during the frequent all-nighters she pulled to get her work out into the world. Getting them into a fashion fit for a golden record, or more common earthbound vinyl, was not easy. The results, however, were worth the effort of working with a system that took up space in multiple rooms.

Even though GROOVE had managed to make it easier to hear the music being made in real time, the entire set up remained complex. Two rooms—a computer room and an analog room—needed to be connected together with a mess of wires, which Don Slepian was hired to keep in order in 1971. This was the laboratory of Max Mathews. Massive amounts of cables made the trip from the digital output of the computer, to the analog equipment in the laboratory. The lab itself was stocked with three reel-to-reel tape recorders, synthesizer modules, and other electronic noise making devices, many of which had been built by Mathews himself. His own workbench was littered with all of the different projects that kept him busy.

Working between two distant rooms had its own unique challenges, even if Spiegel and the others who worked with GROOVE didn't have to wait so long to hear their music. One had to choose either to run the computer and the different devices connected to it, or be in the analog room with hands on the synthesis equipment. Spiegel wrote of the process:

> *The computer sent data for 14 control voltages down to the analog lab over 14 of the long trunk lines. After running it through 14 digital-to-analog converters (which we each somehow chose to calibrate differently), we would set up a patch in the analog room's patch bay, then go back to the computer room and the software we wrote would send data down the cables to the analog room to be used in the analog patch. Many, many long walks between those two rooms were typically part of the process of developing a new patch that integrated well with the controlling computer software we were writing.*

The same back and forth walk applied when it was time to start recording. The musicians could store the time functions they had computed on a "state-of-the-art-washing-machine-sized disk drive." They also had access to a computer tape drive for data storage. When it was time to put sound onto the reel-to-reel audio tape, they would move between the analog room and computer room, making adjustments to each piece of equipment as needed until everything was set and they could essentially hit play.

Every piece on Spiegel's 1980 album, *The Expanding Universe*, was recorded at Bell Labs. She computed in real time the envelopes for individual notes, how they were placed in the stereo field, and their pitches, with Spiegel writing:

> *Above the level of mere parameters of sound were more abstract variables, probability curves, number sequence generators, ordered arrays, specified period function generators, and other such musical parameters as were not, at the time, available to composers on any other means of making music in real time.*

Computer musicians today who are used to working with programs like Reaktor, Pure Data, Max/MSP, Ableton, and Supercollider take for granted the ability to manipulate the sound as it is being made, on the fly, and with a laptop. Back then it was state of the art to be able to do these things, but doing it required huge efforts, and took up a lot of space.

During the height of the progressive rock era, making music with computers was also risky business on the level of personal politics. Computers weren't seen in a positive light. They were the tool of the Establishment, man, used for calculating the path of nuclear missiles and storing your data in an Orwellian nightmare. Musicians who chose to work with technology were often despised. There was an attitude that you were ceding your creative humanity to a cold dead machine. "Back then we were most commonly accused of attempting to completely dehumanize the arts," Spiegel remembers. This macho prog rock attitude haunted Spiegel, despite her being an accomplished classical guitarist and capable of shredding endless riffs on an electrified axe if she chose to.

Laurie Spiegel also took risks inside the avant-garde circles she frequented. Her music was full of harmony when dissonance was all the rage. "It wasn't really considered cool to write tonal music," she said, speaking of the power structures at play in music school. Pieces like the three parts of her "Appalachian Grove" show off her love of folk music and its rhythms, and are evocative of traveling through wooded and mountainous landscapes. The textures are warm and buzzing and there is a beauty and aliveness to the pieces that transcend their digital binary origins. Her "East River Dawn" is the kind of music you would want to listen to after coming home from a hectic day at work. Its relaxing rhythms and gentle arpeggiating tones are entrancing over their fourteen rolling minutes, expansive and hopeful as a new day. The other pieces on her album share the same kind of beauty, exhibiting an ordered world of harmony and resplendent musical geometry.

Vampire

Laurie Spiegel was no stranger to work, and to making the necessary sacrifices so she could achieve her aims and full artistic potential. She supported herself in the '70s in part through soundtrack composition at Spectra Films, Valkhn Films, and the Experimental TV Lab at WNET. TV Lab provided artists with equipment to produce video pieces through an artist-in-residence program. Spiegel held that position in 1976 and composed series music for the TV Lab's weekly show *VTR—Video and Television Review*. She also did the audio sound effects for director David Loxton's sci-fi film *The Lathe of Heaven*, based on the novel by Ursula K. Leguin and produced for PBS by WNET.

Speaking of the Experimental TV Lab, she said, "They had video artists doing really amazing stuff with abstract video and image processing. It was totally different from conventional animation of the hand-drawn or stop-motion action kind. Video was much more fluid and musical as a form."

Spiegel's thinking is multidimensional, and her art multidisciplinary. Working with moving images was a natural extension of her musicality. Between 1974 and 1979 she got the idea that GROOVE could be used to create video art with just a little tweaking of the system. Unlike her *Expanding Universe* album, her video work at Bell didn't get the documentation it deserved. This was in part due to the system's early demise. Hardware changes at the lab prevented many records and tracings from being left behind.

VAMPIRE, however, is still worth mentioning. Standing for "Video And Music Program for Interactive Realtime Exploration/Experimentation," Spiegel was able to turn GROOVE into a VAMPIRE with the help of computer graphics pioneer Ken Knowlton. Knowlton was also an artist and a researcher in the field of evolutionary algorithms, something else Spiegel would later take up and apply to music. In the '60s Knowlton had created BEFLIX (Bell Flicks), a programming language for bitmap computer-produced movies. After Spiegel got to know him, they soon started collaborating. It was another avenue for her to pursue her ideas for making musical structures visible.

Spiegel had reasoned that if computer logic and languages had made it possible to interact with sound in real time, then the GROOVE system should be powerful enough to handle the real time manipulation of graphics and imagery. She started working on this theory first using a program called RTV (Real Time Video) and a routine given to her by Knowlton. She wrote a drawing program, now similar to Microsoft's familiar Paint software, which became the basis on which VAMPIRE was built. Knowlton and Spiegel worked out a routine for a palette of sixty-four definable bitmap textures, which could be used as brushes, alphabet letters, or other images. This was used inside of a box with ten columns, each column having twelve buttons representing a bit that could be on or off. This is how she entered the visual patterns.

In addition to weaving strands of sound, Spiegel was also a hand weaver, and her experience in that art gave her an idea to use with VAMPIRE. Over the years, cards with small holes in them have often been used as one approach to the weaving art form. Card weaving is a way to create patterned woven bands, both beautiful and sturdy, and while some may think that cards are a simple tool, they can produce weavings of infinite design and complexity.

Hand weaving cards are made out of cardboard or cardstock, with holes in them for the threads, very similar to the Hollerith punch cards used for programming computers. From her experience as a weaver she struck upon the idea that she could create punch cards to enter batches of patterns via the card reader on the computer. After she consulted some of her weaving books, she made a large deck of the cards to be able to shuffle and input into the system.

Spiegel quickly found that she enjoyed drawing parameters just like someone would play a musical instrument. Instead of changing pitch, duration, and timbre, she could change the size, color and texture of an image, as she drew it in real time with switches and knobs making it appear on the monitor. Her skills as a guitarist directly translated to this ability, with one hand doing the drawing, as if strumming and plucking the strings, and the other hand changing the parameters of the image using a joystick, as it might change chords on one of her banjos or mandolins.

She saw the objects on the screen as melodies, but it was just one line of music. She wanted more lines, as counterpoint was her favorite musical form, and she wanted to be able to weave multiple strands of images together. She wrote into the program another real time device to interact with. This was a square box of sixteen buttons for typical contrapuntal options as applied to images, giving her a considerable expansion of options and variables to play with.

After all this work, Spiegel eventually hit a wall of what she could achieve with VAMPIRE in terms of improvisation, saying, "The capabilities available to me had gotten to be more than I could sensitively and intelligently control in realtime in one pass to anywhere near the limits of what I felt was their aesthetic potential." It had reached the point where she needed to think of composition.

Ken Knowlton's work with algorithms was beginning to rub off on her and she started to think of, as she put it:

> *the idea of organic or other visual growth processes algorithmically described and controlled with realtime interactive input … It would be possible to compose a single set of functions of time that could be manifest in the human sensory world interchangeably as amplitudes, pitches, stereo sound placements, et cetera, or as image size, location, color, or texture (et cetera), or (conceivably, ultimately) in both sensory modalities at once.*

Ever the night owl, Laurie Spiegel said of her relationship with the system,

> *Like any other vampire, this one consistently got most of its nourishment out of me in the middle of the night, especially just before dawn. It did so from 1974 through 1979, at which time its CORE was dismantled, which was the digital equivalent of having a stake driven through its art.*

Now the tools for making computer music can be owned by many people and used in their own home studios, the echoes of Spiegel's time spent at Bell Laboratories can be found in the work she has done since then, after the death of GROOVE and VAMPIRE. In 1986 she went on to write the Music Mouse software for Macintosh, Amiga, and Atari computers, and also founded the New York University Computer Music Studio. She has continued to write about music for many journals and publications, still composes and has had many of her pieces played and performed.

ALICE in Digital Wonderland

Hal Alles was a hardware engineer working in the field of digital telephony at Bell Labs in the 1970s. The fact that he is remembered as the creator of what some consider the first digital additive synthesizer, and less for his notable contributions to communications, is a quirk of history, as is the fact that he shares a name with the famous computer of *2001: A Space Odyssey*. While other additive synthesizers had been made at Bell Labs, they were done using the MUSIC software, and not as a hardware synth.

Alles hadn't been looking to make a synth while researching methods for echo-cancellation in digital telephone systems. One of the ways he worked around problems with echo was by developing an advanced high-speed oscillator system, which Alles realized might also bear fruit in the world of musical synthesis, as the oscillator's speed leant itself to being used in real time digital control techniques.

Alles needed to sell his digital designs within and without a company that had been the lords of analog, and it needed to be interesting. He came up with a prototype synthesizer to use as a way of demonstrating Bell's digital prowess to his internal and external clients, while entertaining them at the same time.

Alles recalls the atmosphere of those days:

> *As a technology research organization, our research product had a very short shelf life. To have impact, we had to create "demonstrations." We were selling digital design within a company with a 100 year history of analog design... I had developed one of the first programmable digital filters that could be dynamically reconfigured to do all of the end telephone office filtering and tone generation. It*

> *could also be configured to play digitally synthesized music in real time. I developed a demo of the telephone applications (technically impressive but boring to most people), and ended the demo with synthesized music. The music application was almost universally appreciated, and eventually a lot of people came to just hear the music.*

Max Mathews experienced the demo, and was excited by what Alles was doing. Seeing its potential, Matthews encouraged the engineer to develop a digital music instrument, later saying,

> *The goal was to have recording studio sound quality and mixing/processing capabilities, orchestra versatility, and a multitude of proportional human controls such as position sensitive keyboard, slides, knobs, joysticks, etc. It also needed a general purpose computer to configure, control and record everything. The goal included making it self-contained and "portable." I proposed this project to my boss while walking back from lunch. He approved it before we got to our offices.*

Just as Mathews had been able to work on MUSIC during company time, Alles was given the money and time to start working on what would become the Bell Labs Synthesizer, or Alles Machine. It was a 300 pound boat anchor that got the nickname, the Blue Monster, but the name that really stuck was Alice.

Harmonic additive synthesis had already been used back in the 1950s by linguistics researchers working on speech synthesis, and Bell Labs was certainly in on the game. Additive synthesis, at its most basic, works by adding sine waves together to create timbre. Until that time the more common technique had been subtractive synthesis, which used filters to remove or attenuate the timbre of a sound. Computers were already able to do additive synthesis with wavetable oscillators, which had been introduced in MUSIC II, yet additive synthesis could also be done by mixing the output of multiple sine wave generators. This is what Karlheinz Stockhausen did by building up layers of pure sine waves on tape in "Studie I" and "Studie II," rather than with the convenience of a pre-configured synth or computer setup.

Alice was a hybrid machine in that it used a mini-computer to control its complex bank of sound generating oscillators. The mini-computer was an LSI-11, made by the Digital Equipment Corporation, and was one of the earliest 16-bit processors to be made in significant quantities. The computer controlled the sixty-four oscillators, whose output was then mixed to create a number of distinct sounds and voices. The programmable sound generating functions of Alice had the ability to accept a number of different input devices.

It all was connected up to one of Bell Labs' color video monitors. A custom converter was made that sampled the analog inputs and transferred them to 7-bit digital resolution 250 times a second. There were a number of inputs used to work with Alice in real time: two sixty-one-key piano keyboards, seventy-two sliders alongside various switches, and four analog joysticks, just to make sure the user was having fun. These inputs were interpreted by the computer, which in turn controlled the outputs sent to sound generators as parameters. The CPU could handle around 1,000 parameter changes per second before it got bogged down.

The sound generators themselves were quite complex. 1,400 integrated circuits were used in their design. Out of the sixty-four oscillators, the first bank of thirty-two were used as master signals. This meant Alice could be expected to achieve thirty-two-note polyphony. The second set was slaved to the masters and generated a series of harmonics. If this wasn't enough sound to play around with, Alice was also equipped with thirty-two programmable filters and thirty-two amplitude multipliers. With the added bank of 256 envelope generators, Alice had a lot of potential sound paths that could be explored through her circuitry. All of those sounds could be mixed in many different ways into the 192 accumulators she was also equipped with. Each of the accumulators was then sent to one of the four 16-bit output channels then reconverted from digital back into analog on the audio output.

In 1977 the Motion Picture Academy was celebrating the fiftieth anniversary of the talkies. Half a century earlier, the sound work for *The Jazz Singer*, the first talking picture with synced audio and video, had been done by Western Electric with their Vitaphone system. The successful marriage of moving image and sound first seen and heard in that movie wouldn't have been possible without the technology developed by the AT&T subsidiary, and Bell Labs had been invited to be a part of the commemoration of the film for the role they played in syncing sound and image. Alice was chosen as the centerpiece for the event.

For the event, a Bell Labs software junky by the name of Doug Bayer was brought in to improve the operating system of the synth and make the human interface a bit more user friendly. The instrument was flown to Hollywood at considerable risk. The machine was finicky enough without transporting all three hundred pounds of its switches and wires, and it wasn't out of the question that it could get sent into meltdown mode while being banged up on the plane. It was certainly not an instrument for the touring musician.

They hired Laurie Spiegel to be filmed playing Alice. She'd already acquainted herself with the instrument and had learned how to coax out its beautiful sounds. The film of her playing would be shown in the event that Roger Powell, the live musician they hired to play it in Hollywood, wouldn't be able to perform due to malfunction. This film is the only recording of it in live performance left in known existence. Yet to hear how the Bell Labs Digital Synthesizer sounds, look no further than Don Slepian, who was able to make a number of recordings with Alice.

The Digital Bliss of Don Slepian

In 1971 Don Slepian was fresh out of high school and hired by Max Mathews to be a wiring technician for the GROOVE system. As you may recall, Slepian's father had worked with Mathews and Claude Shannon, and had brought home his son a test pressing of *Music from Mathematics*. Now it was young Don's turn to make music of his own, soon becoming a pioneer of ambient music.

Slepian was fascinated by GROOVE but spent most of that summer working on a hardware synthesizer as a gift for John Pierce, who was getting ready to retire from Bell. In an early attempt to synthesize musical timbres from noise sources, Slepian was putting in a lot of time making fixed tuned resonance circuits derived from Max Mathew's electronic violin. Jean-Claude Risset was there doing his residency at the Labs, and after Slepian heard his MUSIC V composition, "Computer Suite from Little Boy," Risset was generous in taking an interest in his own passion for computer music.

In 1972, at age nineteen, Slepian moved to Hawaii, where he taught classes on ARP synthesizers and played as a synthesizer soloist with the Honolulu Symphony. He was, in his own words, "mad for technology in any form," and got involved in the ALOHA project and ARPANET, a project of the Defense Advanced

Research Projects Agency (DARPA), through his father's friends Norman Abramson and Frank Kuo. The ALOHA project was an early computer networking project that used ultra high frequency (UHF) radio frequencies for its operation. Data went over the air as the "transmission of packetized data that helped create the TCP/IP protocol suite." Slepian writes, " I ran a testing outpost that was a teletype and a buffer/transmitter on the third floor of a university building that had a line of sight to the University of Hawaii's computer center. I played text-based games on the early internet when there were only 16 non-classified places to go."

This wasn't his first time around a teletype machine. Back in 1968, when he was fifteen, he had composed a musical piece for three teletypes, "The Chunka Ding Trio." At his high school they had three teletypes—early printers that could be used to send and receive typed messages over telephone and radio—for a minicomputer, whose operation included learning the CTL-G code, which rang the teletype bell, and the CTL-L code, which forced a linefeed. Slepian later recalled his discovery that he could use these two codes to make the machines produce a rhythmic track that sounded like "ding chunka-chunka ding ding chunka-chunka," writing:

> *I cut the slightly oily yellow punched tape where I had recorded these codes, gave it a single twist, and carefully spliced the beginning to the end of the yellow punched tape, creating a loop. I was so pleased with this creation that I created two more variations to this loop and fed it to the other two teletypes. The three teletypes going together would drift against each other, producing a rich ever-changing cacophony of chimes and thunks. When the math teacher who was in charge of the computer room walked in on this impromptu concert he immediately started looking for me. I can't imagine why.*

In 1978, Slepian recorded *Electronic Music from the Rainbow Isle,* and put it out as a self-released cassette. Under the influence of his father, an amateur oboe player who hosted a monthly rehearsal for a woodwind quintet, Slepian grew up listening to Baroque, classical, and Romantic music and began his own piano lessons at age seven. You can hear that classical influence and training all across this work, where it fused with his love for electronics. Tracks like "Horizon" from the album show off the blending of piano, guitar, and electronic textures, while the piece "Glimmerings" revels in the allure of rhythmic, arpeggiated, and sequenced pulses that would later find full and long-form expression on the music he made with the Alles synth.

Slepian had written a number of music scores in MUSIC V using the C programming language under UNIX during the late 1970s, but his main interest was hardware, building filters and oscillators using early operational amplifiers. He fell in love with building new devices, and his felicity with music and circuits was one of the many things that made him a perfect fit for an artist in residency position, which Max Mathews hired him for between 1979 and 1982 to work with the Hal Alles synth.

Mathews wanted Slepian to do something with Alice, which had been resigned to sitting unused in a corner. Alice had become something of an embarrassment due to all the money that had been spent on its creation, quite an achievement in itself considering how much money Bell Labs was accustomed to letting their researchers run with.

Slepian was given a small stipend of $117 a week and a bit of occasional petty cash. He hung out with Mathews and other scientists in the Acoustics and Behavioral Research department and did what he could to make himself useful. He worked on soundtracks for publicity films, and got to see other musicians such as Laurie Spiegel, Roger Powell, and Larry Fast come by and use Alice for projects they were working on. He even showed Bob Moog the machine.

While in Hawaii, Slepian had become entranced by the possibilities of aleatoric composition processes. He had already started using an EML 400/401 sequencer as part of his own synth setups. While a resident at Bell Labs he got to know engineer Greg Sims, who worked with Hal Alles, and convinced him to write a three-stage permutation algorithm using the bottom three programmable sliders on Alice. Slepian recalls,

> *I was in absolute heaven! The Alles Machine had a big digital event sequencer, which was really an array in the LSI-ll's computer memory. I would input 40 or 50 notes into the digital synthesizer using the root, flat 3rd, 4th, 5th, and flat 7th scale degrees. As I would raise the first permutation slider, the sequencer would skip a note and then play a note. Naturally, playing every other note would generate a new melody from my note list.*
>
> *The slider went from 1 to 7. As I raised the slider further to the 5th degree it would play a note, skip five notes and then play the next note. You might think that this scheme would only yield 7 different melodies but that's not true. Since the sequence was playing in real time, the results depended on the exact time when you touched one of the sliders to start the permutation. Now add two more sliders that did the same thing. I could play a note, skip 3, play a note, skip 7, play a note, and then skip 5. With a well-constructed note list I could permutate my sequence all afternoon and not repeat myself. In addition, I could retrograde, transpose, play in dotted rhythms or play in half or double time*

The results of his work are on full display on his 1980 album, *Sea of Bliss*, first released on cassette and featuring two side-length cuts of deep ambient music bringing relaxation and joy to the listener. It's the audio version of taking Valium. Listen to it and feel the stress of life melt away. Don Slepian described his masterpiece for the online *Ambient Music Guide*:

> *It's stochastic sequential permutations (the high bell tones), lots of real time algorithmic work, but who cares? It's pretty music: babies have been born to it, people have died to it, some folks have played it for days continuously. No sequels, no formulas. It was handmade computer music.*

Alice was his everyday music machine for the last two years of its life, and he released a total of five albums featuring its sounds, including his first long playing vinyl record, *Computer Don't Breakdown*. The music features both sequenced and played voices from the synth, and the title forms a musical and computational pun, on the idea of the breakdown in folk music, and what often happens with digital machines.

As the 1980s rolled on, Slepian would continue his dual path as artist and engineer, later working for the Telecommunications for the Handicapped project at BELLCORE, Bell Communications Research, where he built a voice-controlled jukebox with several dozen spoken commands capable of controlling a musical sequence. The machine was trained on the damaged speech of children with cerebral palsy at the Metheny School in New Jersey, after his mother, children's author Jan Slepian, had written the 1980 novel, *The Alfred Summer*, featuring a main character with cerebral palsy. "It was wonderful to see the excitement of the children as they controlled music from their utterances," Slepian says of his jukebox.

From Alice to Amy

After Slepian had done his magic with the machine, the Bell Labs Digital Synthesizer was soon to leave its birthplace. In 1981, Alice was disassembled and donated to the Technology in Music and Related Arts department at the Oberlin Conservatory of Music. Even as a museum piece the synth continued to exert and influence, with a number of commercial synthesizers based on the Alles design produced in the 1980s.

The Atari AMY sound chip is a case in point. Standing for Additive Music sYnthesis and containing sixty-four oscillators reduced down to a single-IC sound chip, AMY had numerous design issues. Digital additive synthesis could now be done with less, but it never really got into the hands of users. Scheduled to be released as part of a new generation of 16-bit Atari computers, and for the next line of game consoles in their arcade division, AMY never saw the light of day. Even after Atari was sold in 1984, she remained waiting in the dark, cut from being included in new products after many rounds at the committee table—that place where so many dreams wind up dead.

Still, other folks in the electronic music industry made use of the principles first demonstrated by Alice. The Italian company Crumar and New York's Music Technologies got into a partnership to create Digital Keyboards. Like Atari, they wanted to resize the Alles Machine, make it smaller, coming up with a two-part invention using a Z-80 microcomputer and a single keyboard with limited controls. They gave it the unimaginative name Crumar General Development System and began selling it in 1980 for $30,000. Since it was way out of the price range of your average musician, they marketed the product to music studios. At that cost, only ten were ever built, but Wendy Carlos got her hands on one, the results of which can be heard on her score to the 1982 film *Tron*.

Other companies got into the game and tried to produce something similar at lower cost, but none of these really managed to find a good home in the market due to the price tag. Then Yamaha released the DX7 in 1983 for $2,000, which implemented FM synthesis and enabled it to achieve many of the same effects as Alice with as few as two oscillators. With the DX7 quickly becoming the first commercially successful digital synthesizer, the demand for additive synths tanked, but Alice's legacy was sealed. What started out as a way for Hal Alles to look at the problem of echoes in digital communications ended up becoming a tool for extending human creativity.

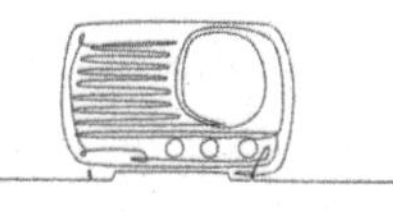

12: FREQUENCY MUTATIONS

John Chowning: Music Hacker

In 1967, FM audio synthesis was discovered by John Chowning during his experiments at Stanford University. FM radio had come along before FM synthesis, but it was Chowning who did the research necessary to apply the pertinent equations to audio, using frequency modulation on audio waveforms in a similar manner to the way frequency modulation is done on radio waves.

John Chowning was born in Salem, New Jersey, in 1934. As a young man he had joined the service and studied music at the Navy School of Music, and next got a bachelor's degree in music in 1959 from Wittenberg University in Ohio. Diploma in hand, he hopped across the pond to study with Nadia Boulanger in Paris, who introduced him to the music of Pierre Boulez and Karlheinz Stockhausen, whose piece "Kontakte" became very attractive to Chowning. For this work Stockhausen invented a unique rotating loudspeaker fitted with a horn attachment. The music from the rotating loudspeaker got picked up by four microphones placed around a circle to create four-channel quadraphonic sound. Chowning was bit by the bug of electroacoustic music and became fascinated with the idea of using loudspeakers in spatialized compositions.

After three years of studies in Paris, Chowning went to Stanford to study composition in 1962, but was disappointed when they didn't have any interest in electronic music. In 1963 Max Mathews wrote his famous paper on the MUSIC IV program, as well as a November 1963 article in *Science* magazine titled, "The Digital Computer as a Musical Instrument." Chowning was given a copy of the article by his friend Joan Mansour, who knew of his electroacoustic experiences in Paris and thought he'd be interested. At the time he hadn't even seen a computer before, yet one of the statements made by Mathews rang inside his head: "There are no theoretical limitations to the performance of the computer as a source of musical sounds, in contrast to the performance of ordinary instruments." Excited by the ideas in the article, Chowning took a computer programming course, convincing him he could learn how to program a computer to make electronic music.

Chowning also had the gumption to get in contact with Mathews and set up a visit to Bell Labs the following summer. While there, Mathews gave him the punch cards that made up the Music IV program from the Bell Labs compiler, which he took back with him to Stanford. He was careful not to drop the box of unnumbered cards, as these only worked when input in the correct order, and programmers still hadn't got into the habit of numbering them. The world of computer music was small back then, still a totally new field. The only other place it was being explored aside from Bell Labs was at Princeton, yet being on the West Coast, isolated from the work of others, became an advantage for Chowning. He was able to pursue computer music without falling under the domineering influence of Milton Babbitt and his theories. "No one cared what was going on at Stanford," Chowning said, allowing him to follow his muse.

Another advantage was access to the state of the art computers at the Stanford Artificial Intelligence Lab (SAIL), where a spirit of interdisciplinary investigation reigned. SAIL had been established in 1962 by professor John McCarthy, a computer and cognitive scientist who founded the field of artificial intelligence, having co-written the paper where that term was first used. McCarthy pioneered computer time-sharing among different users, a solution to the long periods of time needed back when it could take hours for the mainframes to output the intricacies of a program.

With the help of the friendly tuba player and computer whiz David Poole, Chowning got the Music IV program up and running at SAIL. Chowning and Poole used an IBM 1301 disc, which served as the common storage unit between an IBM 7090 and DEC PDP-1 computer, allowing them to save the music they generated. Poole helped Chowning obtain the audio by writing a double buffer program that eventually allowed for two analog signals to be recorded to stereo tape.

Fellow lab rats like Poole provided a hospitable and encouraging environment for learning, and Chowning taught himself the other skills he needed to know as one of the first computer composers and musicians: programming, signal theory, and acoustical physics, all fields of study outside of his initial academic wheelhouse. These new skills opened up further possibilities for Chowning to conduct deep research into the nature and properties of sound and music, enabling him to translate the algorithms for radio frequency modulation (FM) into something that would work for audio frequency modulation.

"Music is a symbolic art," Chowning has said. The Western classical tradition is accustomed to the role of the composer as someone who often puts music into notation before it is ever played by a full ensemble. The music appears first in the imagination, is then composed on paper, and only later performed and heard by musicians. The computer and its programming languages gave composers a different method for musical realization, and, in its capacity as a sonic instrument, gives access to a gamut of timbre that had before only been available in the theoretical realm of ideas. With the tool itself manifested in the physical realm, ideas for compositions followed, and the computer allowed these ideas to be brought down from the platonic realm and realized in a listenable form.

Spatialization, Doppler Shifts, and Vibrato

Like many other composers of his generation who'd been stimulated by the work of Stockhausen, John Chowning became very interested in the spatialization of sound. "Gesang der Jünglinge" was another touchstone, and Chowning set about to conduct his own experiments using a quadraphonic speaker arrangement configured around a listener in the shape of a square. This led him to his discovery of frequency modulation within the range of audio spectra, something he hadn't been looking for, and the subsequent creation of FM synthesis.

One of Chowning's experiments was to divide the levels of intensity between the pairs of left and right speakers to create sound illusions of distance or closeness. Next he worked with Doppler shifts and reverb effects to create the experience of sound moving within what he termed the "listener sound space," an arrangement of speakers with listeners seated within.

Reverb was a key ingredient in his acoustic work and he discovered that if the reverb is applied equally to all channels, it negates any spatialization or perceived distance effects of the audio. From this he learned there are two types of reverb: global and local. The global is applied to all sounds equally in a mix, while the local is applied only to certain signals emerging from specific loudspeakers. Chowning then came up with an equation that showed how reverberation within a small space remains basically constant, even as signal distance is increased. In a large space, the equation can determine the distance of a sound based on the ratio of reverbant and non-reverberant signal.

With equations in hand, Chowning programmed a spatialization routine with the MUSIC program in 1972, featuring a graphical aspect that allowed the composer to draw the trajectory of sound movement from one speaker to another. This program had two different aspects of velocity that could be used: angular and radial. The angular velocity is the rate of change of the sound intensity, whereas the radial velocity represents the rate of frequency shift in sound, also known as the Doppler effect.

Best demonstrated by the blaring sirens of an emergency vehicle, the Doppler effect is most often heard in everyday life as the sounds of objects moving closer toward a listener, and then farther away. The Doppler effect can also be experienced when there is a loud stationary source of sound, but the listener is moving around it, such as the bump of bass emanating from a house party on a Friday night while a couple walks their dog around the block where the party is taking place.

The Doppler shift was first discovered in the light spectrum by Austrian physicist Christian Doppler, who wrote of the phenomenon in his 1842 paper, "On the coloured light of the binary stars and some other stars of the heavens." Three years later, Buys Ballot ran tests to see if Doppler shift was also present in the audio spectrum, showing that it was. He was able to show that the pitch of a sound is higher than the emitted frequency as the sound source approaches, and becomes lower than the emitted frequency when it recedes. Further, the French physicist Hippolyte Fizeau independently discovered the property within electromagnetic waves in 1848. Since that time, a number of equations came into use to mathematically model the phenomenon. The Doppler effect has gone on to be used in a number of settings, such as radar, satellite navigation and communication, medicine, astronomy and the ubiquitous use of sirens.

In working with sound intensity, Doppler effects, and reverberation, Chowning realized there was much more going on in the perception of loudness in space than simply the distance and decay rate of audio as it travels. Vibrato was another factor in acoustics that could change the way a sound was perceived, providing the next key he needed to unlock audio FM synthesis.

The computer generated waveforms Chowning created were not natural. In nature, sounds are quasi-periodic, yet a computer is capable of making a perfect periodic sound, or a soundwave that is repeated exactly multiple times. Some critics of computer music have pointed out the unnatural sound generated by these electrons. To make the timbres sound more natural, variations have to be created in the waveform to make them quasi-periodic, or less than perfect. Chowning did this by micro-modulating the frequency with vibrato.

This led to two discoveries. For one, when a sound is made of multiple partials, he realized that adding small but equal amounts of vibrato to each partial creates perceptual fusion, or the illusion in the listener that the sound is one single tone. Perceptual fusion is also at work in film, where the eye thinks all the motion is one continuous whole, when in reality it's a sequence and series of projected frames. His second discovery was source aggregation. This can be created when small non-equal amounts of vibrato are applied to groups of partials, which the listener perceives as separate tones and sounds.

The Birth of FM Synthesis

The same principle at work in radio FM, where a carrier signal is modulated by the input signal, is used in FM synthesis. Audio FM synthesis is achieved by using one signal, called the modulator, to change the pitch of another signal, called the carrier, within a similar audio range. This modulation adds new information to the carrier signal and changes its timbre. The use of multiple modulators on one carrier gives further variables for shaping the final sound signal.

The stage had been set for this discovery as Chowning continued to explore the effects of vibrato. He noticed that when the rate of vibrato entered the audio range at twenty hertz, partials started to form within the spectrum. He also noted how the relationship between the modulator and the carrier determined whether a sound was harmonic or inharmonic, as well as producing changes in the timbre.

Chowning soon learned that if the modulator frequency is a whole number multiple of the carrier frequency, then the partials will be harmonic. Next he discovered the modulation index—the ratio between the depth of modulation and modulation frequency—learning this could be used to change a signal's bandwidth over time when the amplitude envelope of an entire signal is added to the value of the modulation index. This creates extra audible partials to change the sound.

Similar effects had been achieved with additive synthesis, but those often require up to sixteen or more oscillators, whereas FM synthesis could achieve great results with two oscillators, the modulator and the carrier, though more can also be used.

FM Mutations

In the summer of 1967, John Chowning visited Jean-Claude Risset and Max Mathews at the Bell Laboratories, before making his discovery a few months later in the fall. In December he visited Bell Labs again, and Risset took notes about this innovation in synthesis. Risset recalled, "He explained his ongoing experiments on illusory moving sound sources and on spectral change through high-speed frequency modulation. He gave me his data together with a tape, so I could use the process right away."

From his notes, Risset did his own work and ended up using FM synthesis in conjunction with other computer techniques to create "Mutations" in 1969, premiering at the Moderna Museet in Stockholm in 1970. Commissioned by GRM and composed entirely on computer and two-track tape at Bell Labs, it explores the idea of composing at the very level of sound itself. Gradual changes or mutations occur over the course of the piece, "including the shift from a range of discontinuous heights to continuous frequency variations." Familiar bell and organ tones get warped in a fun house mirror of auditory illusion across the span of the piece. One section conjures up images of particle showers raining down from space, each sound a discrete electron. The piece is famous for its use of the endless glissando, or barber pole of sound, which Risset had devised for his previous composition, "Computer Suite from Little Boy," in 1968. The musical barber pole was a variant of the Shepard tones also created at Bell Labs using MUSIC software. The ending section drifts along with the tinkle of bells and electronic wind that sounds as if it is blowing grains of sand across a vast desert.

Sabelithe and Turenas

Whereas Rissett's "Mutations" had used FM synthesis as one of many components, Chowning's work "Sabelithe" was the first realized purely with FM synthesis alone. An early version of the work was first composed in 1966, but it was never completed because of a move of SAIL to the DC Power Laboratory building, with Chowning revising the work for quadraphonic tape in 1971. It was his first computer piece and made use of his discovery and spatialized sound. Gentle tinklings and short tones transform into an ascending coruscation, and descend back into low sounds and guttural clicks, before ascending again into complex crystalline geometric structures.

Chowning's next composition, 1972's "Turenas," makes use of FM synthesis, his surround sound setup, and programming for Doppler shifts in MUSIC IV. The title itself is an anagram of "natures," and references Chowning's intention to create realistic timbral sounds with artificial means. The first of the piece's three movements makes use of the mathematical formula for a Lissajous pattern, also called a Bowditch curve, produced when two sinusoidal curves intersect with their axis at right angles to each other. It was first studied in 1815 by American mathematician Nathaniel Bowditch, while the curves were later studied by Jules-Antoine Lissajous, a mathematician from France who used a compound pendulum that poured out narrow streams of sand to study the pattern.

The curve is well known in the world of electronics, where it can be made visible using an oscilloscope. With the oscilloscope, the shape of the curve shows characteristics of electronic signals, used to study the properties of any pair of simple harmonic motions at right angles to each other. The Lissajous patterns came to be used in determining the frequencies of sounds or radio signals. A known signal frequency is put onto the horizontal axis of an oscilloscope, and the signal that needs to be measured is put on the vertical axis. The pattern that results is a function of the ratio of the two frequencies.

When Chowning had originally made a sketch of the proposed movement of sound in space for "Turenas," an engineer at Stanford commented that it looked like the Lissajous pattern, inspiring Chowning to go ahead and make implicit use of the pattern. One of the properties of a Lissajous path is that its rate of change slows down as it reaches its peak, like a car set on cruise control at seventy miles per hour.

Chowning used a double Lissajous to surround the listener in these mathematical patterns. The second movement is a tour de force of everything Chowning had learned. He uses reverberation, vibrato, modulation, and many timbral transformations to showcase the veracity of FM synthesis. One way to hear the composition is as an artificial lifeform recalling the birth of these methods in the womb of SAIL. Chowning's luminous silicon life form expands as a fractal pattern in space, transforming first into the long sonorous tone of a bell, and back again into microcosmic oscillations. These again mutate into brass-like tones that again shift into a pitter patter of sputtering that presage the clicks and cuts of the IDM style of electronica to emerge in the 1990s. Like nature itself "Turenas" is constantly mutating, one part becoming something else just before it can stabilize in one recognizable form, moving on to become something else again, shifting in time as it moves about in space.

Stria and the Golden Ratio

Johannes Kepler, the great astronomer and explorer of the harmony of the spheres, once said, "Geometry has two great treasures: one is the theorem of Pythagoras, the other the division of a line into mean and extreme ratios, that is Phi, the Golden Mean. The first way may be compared to a measure of gold, the second to a precious jewel."

The Golden Mean, often also called the Golden Ratio or Golden Section, has been studied since at least the time of Euclid. It is commonly symbolized by the Greek letter Phi (φ) , giving it another moniker, the Phi Ratio. The Golden Mean can be found when a line is divided into two, so that the whole length divided by the long part is also equal to the long part divided by the short part. In math, two numbers are in the Golden Mean if the ratio of the sum of the numbers (x + y) divided by the larger number, x, is equal to the ratio of the larger number divided by the smaller number (x/y). Phi is an irrational number equal to 1.6181.618033988749.... and then continues on and on like the number Pi.

The Golden Mean can be found in the sacred art and architecture of many traditional civilizations, from Egypt to Islam, from China to the Gothic cathedrals of the Middle Ages, and many points in between. It can also be found in many natural forms, such as certain leaves and the shell of the chambered nautilus. Wherever it is found, there exists a manifestation of this natural harmony.

Chowning's most famous work, 1977's "Stria," adheres with rigor to the use of the Golden Ratio in all parameters and aspects of the composition. It also makes strict use of FM synthesis. In what can be heard as a response to Goethe, who once said, "Geometry is frozen music," Chowning took the sacred proportions of the Golden Ratio and unfroze them so that they could be listened to.

In his FM research, Chowning discovered that when he composed using powers of the Golden Ratio, applying them to the carrier-to-modulator frequency, sideband components, or portions of audio spectrum above and below the carrier, were obtained that were also powers of the Golden Ratio. He decided to base the macrostructure and the microstructure of "Stria" on the Golden Ratio where everything revolves around the number 1.618. The first frequency heard in the piece is 1,618 hertz, and all the durations in the piece relate to Phi.

"Stria" was written using MUSIC 10 at SAIL, and travels from highs to lows as it traverses the mathematics of the Golden Ratio in different ways. The precise use of computer controlled timbre and vibrato throughout give the piece a sound that is artificial, yet also natural sounding because of the Phi ratio as the structural component. Listening to it is like receiving a geometric download from the platonic realm.

The Center for Computer Research and Musical Acoustics

Between 1966 and 1967 Karlheinz Stockhausen had come to lecture at the University of Davis in California, and while he was on the West Coast the maestro visited John Chowning at Stanford to see and hear the work he was doing with quadraphonics and the creation of illusory space. He was struck by the importance of Chownings work, and realized it was a powerful step in the evolution of electronic composition. One of his questions was, if he learned how to work with a computer, would he be able to get more pieces done? Chowning said, "I doubt it, because it takes a long time and a lot of care." He thought that in Stockhausen's instance, with his sensitive and fine tuned sense of hearing, the long time needed for programming wouldn't be expedient. Even so, when Stockhausen went back to Europe, he told his friend Pierre Boulez to pay attention to the work being done at Stanford.

The computer was becoming an important tool and it would do well to get it into the hands of more composers and musicians. In 1969, the first summer workshop on computer generated music was taught at Stanford by Chowning, Max Mathews, musician and computer scientist Leland Smith, and engineer George Gucker. This became the seed for the creation of the Center for Computer Research in Music and Acoustics (CCRMA).

CCRMA was officially founded at Stanford in 1974 by Chowning, Smith, and composer Loren Rush. Other founding members included James Andy Moorer, a digital audio and computer engineer, and John M. Grey, who were working on the analysis and synthesis of acoustic instrument sounds by computer. John Pierce later came to work at CCRMA after leaving Bell Labs.

The CCRMA continued, and continues still today, to have significant impacts in the arts and sciences as a multi-disciplinary facility where composers and researchers work together, exploring everything from psychoacoustics and data sonification to perceptual audio coding and beyond. At the same time they were getting up and running, IRCAM was in its planning stages, with Pierre Boulez relying heavily on the groundwork they laid in designing his own laboratory of sound.

The DX7

If all of this sounds a touch esoteric, then let it not be forgotten that John Chowning made a lasting imprint on the popular music of the 1980s and onwards when Yamaha licensed the technology of FM synthesis to create their DX7 synthesizer—regarded as the first commercially successful digital synthesizer, not to mention one of the best selling synths of all time. Work on its development began in 1974, but a commercial synth wasn't available until 1983. With its ability to imitate acoustic instruments like piano, brass, and woodwind, as well as create new timbres distinct from earlier analog synths, it quickly became a hit with musicians when it was released.

The DX7 was hard to program through its complex menus. As a result, many who worked with it used its out-of-the-box presets, and these sounds became staples in '80s music. The electric piano sound alone was used on more than 40 percent of the top hits from 1986. Brian Eno got one and it became a crucial part of the setup and workflow in his home studio, with the famed producer noting:

> *I use the DX7 because I understand it. I was quite ill for a while, and I filled the time by learning it. I think it's just as good as anything else. Sticking with this is choosing rapport over options. I know that there are theoretically better synths, but I don't know how to use them. I know how to use this. I have a relationship with it.*

The DX7 is programmed with thirty-two sound-generating algorithms, each a different arrangement of its six sine wave operators. These give the DX7 its classic bright and glassy sound, as heard on a-ha's "Take on Me" and Kenny Loggins' "Danger Zone," both of which rely on the synth's BASS 1 preset. The keyboard itself spans five octaves, and has sixteen-note polyphony. New patches can be created within its deep menu system rather than the cables required for analog synths, and these patches could be named and saved inside its memory bank.

John Chowning also thought the DX7 could be used to teach about the properties of sound, saying, "Many basic acoustic phenomena can be demonstrated quite easily using the DX7. It could become an incredibly powerful tool for learning acoustics and psycho-acoustics at a very simple level."

After the success of the DX7, which has sold over 200,000 units since release, Yamaha released a plethora of lower cost FM synthesizers. A cheap version of the DX7 soundchip also went into the Sega Genesis, making it the sound a generation of video game heads grew up jamming their thumbs too.

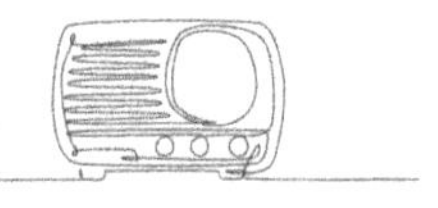

13: SPEAKINGS

The Musical Poetics of Pierre Boulez

Pierre Boulez was of the opinion that music is like a labyrinth, a network of possibilities that can be traversed by many different paths, not needing to have a clearly defined beginning, middle, and end. Like the music he wrote, the life of Boulez did not follow a single track, but shifted according to the choices available. After all, not all of life is predetermined. Even if the path of fate has already been cast, choices remain open, and Boulez held that music is an exploration of these choices.

In an avant-garde composition a piece might be tied together by rhythms, tone rows, and timbre, just as a life might be tied together by relationships, jobs, works made, and things done. As Boulez wrote:

> *A composition is no longer a consciously directed construction moving from a "beginning" to an "end" and passing from one to another. Frontiers have been deliberately "anaesthetized," listening time is no longer directional but time-bubbles, as it were...A work thought of as a circuit, neither closed nor resolved, needs a corresponding non-homogenous time that can expand or condense.*

Boulez was born in Montbrison, France, on March 26, 1925, to an engineer father. As a child he took piano lessons, played chamber music with local amateurs, and sang in the school choir. Boulez was gifted at mathematics and his father hoped he would follow him into engineering, beginning an education at the École Polytechnique, but music intervened. His world was rocked after seeing the operas *Boris Godunov* and *Die Meistersinger von Nürnberg*. When he next met the celebrity soprano Ninon Vallin, the two hit it off and she asked him to play for her, helping persuade his father to let him apply to the Conservatoire de Lyon. He auditioned twice and didn't make the cut, but this only furthered his resolve to pursue a life path in music. A door opened when he was admitted to the preparatory harmony class of composer Georges Dandelot, after which his ascension into the world of music was swift.

Two of the choices Boulez made that had a long-lasting impact on his career were in his choice of teachers, the first being Olivier Messiaen, who he approached in June 1944. Messiaen taught harmony outside the bounds of traditional notions, and embraced the new music of Schoenberg, Webern, Bartok, Debussy, and Stravinsky. In February of the following year, Boulez attended a private performance of Schoenberg's "Wind Quartet" conducted by René Leibowitz, and the event left him breathless, leading him to his second influential teacher. Leibowitz had studied with Schoenberg and Anton Webern, and was a friend of Jean-Paul Sartre, with his performances of music from the Second Viennese School making him something of a rock star in avant-garde circles of the time.

Under the tutelage of Leibowitz, who Boulez organized a group of students to take lessons from, he was able to drink from the fountain of twelve-tone theory and practice. Boulez later told *Opera News* that this music "was a revelation," calling it, "a music for our time, a language with unlimited possibilities. No other language was possible. It was the most radical revolution since Monteverdi. Suddenly, all our familiar notions were abolished. Music moved out of the world of Newton and into the world of Einstein."

In 1946, the first public performances of Boulez's compositions were given by pianist Yvette Grimaud. He kept himself busy living the art life, tutoring the son of his landlord in math to help make ends meet. He made further money playing the ondes Martenot, an early French electronic instrument designed by Maurice Martentot who had been inspired by the accidental sound of overlapping oscillators heard while working with military radios. Martentot wanted his instrument to mimic a cello, and Messiaen had used it in his famous "Turangalîla-Symphonie," written between 1946 and 1948. Boulez got a chance to improvise on the ondes Martentot as an accompanist to radio dramas.

In 1949, Boulez met John Cage when he came to Paris and helped arrange a private concert of the American's "Sonatas and Interludes for Prepared Piano." Afterwards, the two began an intense correspondence that lasted for six years. In 1951 Pierre Schaeffer hosted the first musique concrète workshop. Boulez, Jean Barraqué, Yvette Grimaud, André Hodeir, and Monique Rollin all attended, while Olivier Messiaen was assisted by Pierre Henry in creating a rhythmical work, "Timbres-durè es," made from a collection of percussive sounds and short snippets.

At the end of 1951, while on tour with the Renaud-Barrault company, Boulez visited New York for the first time, staying in John Cage's apartment and was introduced to Igor Stravinksy and Edgard Varèse. Cage was becoming more and more committed to chance operations in his work, and this was something Boulez could never get behind. Instead of adopting a "compose and let compose" attitude, Boulez withdrew from Cage, and later broke off their friendship completely

In 1952, Boulez met Stockhausen who had come to study with Messiaen, and the pair hit it off, even though neither spoke the other's language. Their friendship continued as both worked on pieces of musique concrète at the GRM, with Boulez's contribution being his "Deux Études," a mechanical sounding work of grinding gears, turbulent dissonance, and fanciful transformations of percussive objects. In turn, Boulez went to Germany in July for the summer courses at Darmstadt. There he met Luciano Berio, Luigi Nono, and Henri Pousseur, among others, and found himself moving into a role as an acerbic ambassador for the avant-garde.

Sound, Word, Synthesis

As Pierre Boulez got his bearings as a young composer, the connections between music and poetry captured his attention, as they had Schoenberg. Poetry became integral to Boulez's orientation towards music, and his teacher Messiaen would say that the work of his student was best understood as that of a poet.

"Sprechgesang," or speech song, a kind of vocal technique that straddles the line between speaking and singing, was first used in formal music by Engelbert Humperdink in his 1897 melodrama *Königskinder*. In some ways, sprechgesang is a German synonym for the already established practice of the "recitative" in operas, as found in Wagner's compositions. Arnold Schoenberg used the related term "sprechstimme" as a technique in his 1912 song cycle, "Pierrot Lunaire," where he employed a special notation to indicate the parts that should be sung-spoke. Schoenberg's disciple Alban Berg used the technique in his opera *Wozzeck* (1924), before Schoenberg employed it again in his opera *Moses and Aron* (1932).

In Boulez's explorations of the relationship between poetry and music, he questioned "whether it is actually possible to speak according to a notation devised for singing. This was the real problem at the root of all the controversies. Schoenberg's own remarks on the subject are not in fact clear."

Pierre Boulez wrote three settings, or musical adaptations, of René Char's poems, "Le Soleil des eaux," "Le Visage Nuptial," and "Le Marteau sans maître." Char had been involved with the Surrealist movement, was active in the French Resistance, and mixed freely with other Parisian artists and intellectuals. "Le Visage Nuptial" ("The Nuptial Face") from 1946 was an early attempt at reuniting poetry and music from the diverging paths they had taken so long ago. In it, Boulez took five of Char's erotic texts and wrote the piece for two voices, two ondes Martenots, a piano, and percussion. In the score there are instructions for "Modifications de l'intonation vocale."

In a personal artistic breakthrough, his next attempt in this vein was "Le Marteau sans maître" ("The Hammer without a Master") composed between 1953 and 1957, and it remains one of Boulez's most regarded works. He brought his studies of Asian and African music to bear on the serialist vortex that had sucked him in, thereby spitting out one of the stars of his own universe.

The work is made up of four interwoven cycles with vocals, each based on a setting of three poems by Char taken from his collection of the same name, and five of purely instrumental music, the wordless sections acting as commentaries to the parts employing sprechstimme. First written in 1953 and 1954, Boulez revised the order of the movements in 1955, while infusing it with newly composed parts. This version premiered that year at the Festival of the International Society for Contemporary Music in Baden Baden. Once finished, Boulez had a hard time letting his compositions just be, and tinkered with it some more, creating another version in 1957.

"Le Marteau sans maître" is often compared with Schoenberg's "Pierrot Lunaire." By using sprechstimme as one of the components of the piece, Boulez is able to emulate his idol while contrasting his own music from that of the originator of the twelve-tone system. As with much music written by his friends Cage and Stockhausen, the work is challenging to the players, with the biggest challenges directed at the vocalist, whose directions include humming, glissandi, and jumping over wide ranges of notes.

The work embodies Char's idea of a "verbal archipelago," where the images conjured by the words are like islands that float in an ocean of relation, but with spaces between them. The islands share similarities and are connected to one another, but each is also distinct and of itself. Boulez took this concept and created his work where the poetic sections act as islands within the musical ocean.

A few years later, he worked with material written by the symbolist and hermetic poet Stéphane Mallarmé, when he composed "Pli selon pli" in 1962. In particular it takes influence from Mallarmé's poem "A Throw of the Dice," in which the words are placed in various configurations across the page, with changes of font size and instances of italics or all capital letters. Boulez took these and made them correspond to changes to the pitch and volume. The title, meanwhile, comes from a different poem by Mallarmé, "Remémoration d'amis belges," in which he describes a mist gradually covering the city of Bruges until it disappears.

Subtitled "A Portrait of Mallarmé," Boulez uses five of Mallarmé's poems in chronological order, starting with "Don du poème" from 1865 for the first movement and finishing with "Tombeau" from 1897 for the last. Some consider the last word of the piece, *mort*, or *death*, to be the only intelligible word in the work. The voice is used more for its timbral qualities, and to weave in as part of the course of the music, than as something to be focused on alone.

Later still, in 1970, Boulez took E.E. Cummings poems and used them as inspiration for his work "Cummings Ist der Dichter." Needless to say, Boulez worked hard to relate poetry and music together in his work. It is no surprise, then, that the institute he founded would go far in giving machines the ability to sing, and in fostering the work of other artists who were interested in the relationships between speech and song.

Ambassador of the Avant-Garde

At the end of the 1950s, Boulez had left Paris for the West German town of Baden-Baden, where he had scored a gig as composer in residence with the Southwest German Radio Orchestra. Part of his work consisted of conducting smaller concerts, while giving him access to an electronic studio where he set to work on a new piece, "Poesie Pour Pouvoir," for tape and three orchestras. Baden-Baden would become his home, eventually buying a villa there, a place of refuge to return to after his various engagements that took him around the world and on extended stays in London and New York.

In the 1970s, Boulez had a triple coup in his career. The first part of his tripartite attack for avant-garde domination involved his becoming conductor and musical director of the BBC Symphony Orchestra. The second part came when Boulez was offered the opportunity to replace Leonard Bernstein as conductor of the New York Philharmonic. He felt that through innovative programming, he would be able to remold the minds of music goers in both London and New York. The third prong came when he was asked by the president of France to come back to his home country and set up a musical research center.

Institute for Research and Coordination in Acoustics/Music

Back in 1966, Pierre Boulez had proposed a total reorganization of French musical life to André Malraux, the Minister of Culture. Malraux rebuffed Boulez when he appointed Marcel Landowski, who was much more conservative in his tastes and programs, as head of music at the Ministry of Culture. Boulez, who had been known for his tendency to express himself as an epic jerk, was outraged. In an article he wrote for the French magazine *Nouvel Observateur,* he announced that he was "going on strike with regard to any aspect of official music in France."

When confronted about this aspect of his reputation later in life, Boulez said, "Certainly I was a bully. I'm not ashamed of it at all. The hostility of the establishment to what you were able to do in the Forties and Fifties was very strong. Sometimes you have to fight against your society."

So when Boulez was asked by French President Georges Pompidou to set up an institute dedicated to researching acoustics, music, and computer technology, he was quick to recant his strike with regards to official music in France and get busy with work. This was the beginning of the Institut de recherche et coordination acoustique/musique, or the Institute for Research and Coordination in Acoustics/Music (IRCAM). The space was built next to, and linked institutionally to, the Centre

George Pomidou cultural complex in Paris, and official work started in 1973.

Boulez modeled the institute after the Bauhaus, the famed interdisciplinary school of art in Germany that provided a meeting ground for artists and scientists from 1919 to 1933. His vision for the institute was to bring together musicians, composers, scientists, and developers of technology. In a publicity piece for IRCAM he wrote:

> *The creator's intuition alone is powerless to provide a comprehensive translation of musical invention. It is thus necessary for him to collaborate with the scientific research worker... The musician must assimilate a certain scientific knowledge, making it an integral part of his creative imagination...at educational meetings scientists and musician's will become familiar with one another's point of view and approach. In this way we hope to forge a kind of common language that hardly exists at present.*

To bring his vision into reality, he needed the help of those at the forefront of computer music. To that end, Boulez brought Max Mathews on board as a scientific advisor to the IRCAM project, and he served in that capacity for six years between 1974 and 1980. Mathews' old friend Jean-Claude Risset was hired to direct IRCAM's computer department, which he did between 1975 and 1979. The work that their colleague John Chowning was doing back in California was also crucial to the success of the institute, and he was tapped as a further resource.

Putting together IRCAM was a project that went on for almost a decade before it was fully up and running, and from 1970 to 1977 most of the work done was the preliminary planning, organization, and building of the vessel that would house the musical laboratory. Unlike the BBC or the West German Radio, it did not have the advantage of being part of an existing institution, so everything, including the space, had to be built from scratch. There were several existing templates for electronic music and research that IRCAM could have followed, and it chose the American template, modeled on the work done at Bell Labs and the CCRMA, when Max Mathews was asked to be the scientific director of IRCAM in 1975.

In 1975 Pierre Boulez spent two weeks at the CCRMA, studying all they were getting up to and forging a lasting connection with IRCAM. One of the results was that a lot of the American computer workers helped set up IRCAM's initial system until the French had enough people trained in the technology themselves. Working with the same systems meant those used at each institution were compatible with each other, enabling extensive back and forth visitations between CCRMA and IRCAM staff. James Moorer did a residence at IRCAM, and John Chowning went on to become a guest artist there on multiple occasions. Chowning's "Phoné" is a case in point of the cross collaboration between the two institutes.

Phoné

Chowning had been inspired by the work that Michael McNabb was doing with additive synthesis of vocals at the CCRMA, but he himself hadn't planned on working on synthesizing the voice. Then, when Chowning went to work at IRCAM at the invitation of Jean-Claude Risset, he thought it might be a good project to try to build up believable singing voices using the FM technique. Then he ran into Johan Sundberg, a Swedish scientist and musicologist who had done research on organ pipes and analysis of the singing voice. Chowning proceeded to pick Sundberg's brain and, by his own description, became "seduced by the problem" of making computers sing using FM synthesis. This in turn led to the creation of algorithms used for vocal synthesis at IRCAM using one of their DEC PDP-10 computers. When Pierre Boulez first heard what Chowning was doing, he thought Chowning must have sampled a human voice, because the sound was so clear. Chowning had to assure him that it was all done with FM synthesis algorithms.

Back home in the States, Chowning coded the piece at CCRMA between 1980 and 1981 using the Samson Box, a real time computer audio processor designed by Peter Samson that was much faster than other systems at the time. To deliver the goods on mimicking human speech, a number of characteristics particular to the voice needed to be implemented, and these marked difficult technical hurdles and fine detail in the attack of the sound and vibrato. He worked on all of these and other little details for about six months, until they became believable. The FM techniques he was employing offered an advantage over linear predictive coding and other methods of voice modeling, which were expensive to compute. "FM was incredibly fast and very easily controlled," Chowning said in relation to the different computing methods.

"Phoné" is the ancient Greek word for "voice," which of course is the same word used to denote one of the main tools in telecommunications. Using FM synthesis, Chowning was able to transform the voice of the bell into a number of different timbres, including that of a human voice with simulated formants. He plays with perception by mixing in sounds that are possible without the aid of a computer with sounds that could not be created any other way, and further made a computer sing beyond the physical limitations of even the most highly trained human singer. To create the sound, he divided the pitch space into pseudo octaves using the Golden Ratio instead of powers of two.

"Phoné" was premiered at the Centre Pompidou by IRCAM in February of 1981. Chowning would use his pitch structure derived from the Golden Ratio again in his 2005 work "Voices," for soprano and interactive computer.

Intercontemporary Underground Music

Much of the space for IRCAM was built below ground, beneath the Place Igor Stravinsky in Paris, where the boisterous noise of the city streets above wouldn't penetrate. The underground laboratories were first inaugurated 1978 and contained eight recording studios, eight laboratories, and an anechoic chamber, plus various offices and department spaces. Though it has since been reorganized with the passing of the years, it was first arranged into five departments, each under its own composer-director, with Boulez as the tutelary head. These departments were Electro-Acoustics, Pedagogy, Computers, Instruments, and Voice, as well as a department called Diagonal that coordinated between the other departments that largely followed their own research and creative interests.

The piece de resistance at IRCAM is the large Espace de Projection, also known as Espro, a modular concert hall whose acoustics can be changed according to the temperament and design of the composers and musicians working there. The Espro space was created under the direction of Boulez and features a system of "boxes in boxes" to create the variable acoustics. When the space was first opened, Boulez said it was "really not a concert hall, but it can project sound, light, audiovisual events, all possible events that are not necessarily related to traditional instruments." The position of the ceilings can be moved to change the volume of the room, and the walls and ceilings have panels that are made of rotatable prismatic modules that each have three faces, one for absorbing, another for reflecting, and one for diffusing sound. These so-called "periacts" can be changed on the spot.

Pierre Boulez was busy as all get-out in the 1970s. If it wasn't enough to be developing IRCAM, conducting the BBC Symphony Orchestra from 1971 to 1975, and conducting the New York Philharmonic from 1971 to 1978, he also founded the Ensemble intercontemporain (EIC) in 1976. The EIC was built up with support from French Minister of Culture Michel Guy and the British arts administrator Nicholas Snowman. Boulez wanted to cultivate a group of musicians dedicated to performing contemporary music, and EIC would have a strong working relationship with IRCAM, so that musicians were available to play compositions made in conjunction with the institute inside the Espro, as well as tour and make recordings. This of course included Boulez's own compositions as he had the energy to return to writing music as his conducting activities slowed down.

Though Boulez had made a piece of musique concrète at GRM, and had experimented with tape music with "Poesie Pour Pouvoir," these were not his main interests in avant-garde music. What concerned Boulez was the live transformation of acoustic sound electronically. He felt that recordings, played in a concert hall, were like going to listen to a dead piece of music. The transformation of live sound was what held promise in his mind. While the possibility for the live transformation of acoustic sounds had been explored by Stockhausen and Cage, these did not have the same precision that was now available with the computers and programs created at CCRMA and IRCAM.

"Répons" was created in various versions between 1980 and 1984. The instrumental ensemble is placed in the middle of the hall, while six soloists are placed at various points around the audience, with the six loudspeakers facing the listeners. The solo instruments include two pianos, harp, cimbalom, vibraphone, and glockenspiel or xylophone, and it is these instruments that give "Répons" much of its color. Boulez has said that the title of the work is a portmanteau of words whose meaning is dialogue and response, which indicates the way the instrumental music gets transformed by computers, which take the acoustic music and change it with effects or other treatments and project it through the performance space via the loudspeakers. In "Répons," the harp, vibraphone, and piano create glittering sparkles that illuminate the space, fulfilling Boulez's dream of the live electronic metamorphosis of acoustic sound.

Max/MSP

Boulez's vision of a "general school or laboratory" where scientists and sound artists mixed and mingled had come to fruition, and once IRCAM got into a groove, it started pushing out a steady stream of compositions, papers, and software from its many scientific and artistic residents and collaborators. One of its most famous outputs is the software suite Max/MSP.

Building off the gains in musical software developed by Mathews, Miller Smith Puckette (MSP) started to work on a program originally called "The Patcher" at IRCAM in 1985. The first version for Macintosh had a graphical interface that allowed users to create interactive scores, but it wasn't yet powerful enough to do real time synthesis, instead using MIDI and similar protocols to send commands to external sound hardware.

Four years later, Max/FTS (Faster Than Sound) was developed at IRCAM, named in honor of Mathews. This version could be ported to the IRCAM Signal Processing Workstation (ISPW) for the NeXT computer system. This time around it could do real time synthesis using an internal hardware digital signal processor (DSP), making it a forerunner to the MSP extensions that would later be added to Max. In 1989, the software was licensed to Opcode, who promptly launched a commercial version at the beginning of the next decade.

Opcode held onto the program until 1997. During those years a talented console jockey named David Zicarelli further extended and developed the promise of Max. Though Opcode wanted to cancel their run with the software, Zicarelli knew it had even further potential, so he acquired the rights and started Cycling '74. Zicarelli's timing proved to be fortuitous, as Gibson Guitar ended up buying Opcode and closing it a year later. Such is the fabulous world of silicon corporate buyouts.

In the meantime, Miller Smith Puckette had released the independent and open-source composition tool Pure Data (Pd). It was a fully redesigned tool that still fell within the same tradition as his earlier program for IRCAM. Zicarelli, sensing that a fruitful fusion could be manifested, released Max/MSP in 1997, the MSP portion being derived from Puckette's work on PureData. The two have been inseparable ever since.

The achievement meant that Max was now capable of real time manipulation of digital audio signals without dedicated DSP hardware. The reworked version of the program was also something that could work on a personal computer or laptop, finally allowing composers to use this powerful tool in their home studios. At long last, the musical composition software that had begun on extensive and expensive mainframes, and had required cultural connections to work at places like Bell Labs or IRCAM, was now available to those who were willing to pay the entrance fee. And if you had a computer but couldn't afford the commercial Max/MSP, you could still download Pd for free. The same is true today.

Extension packs were now being written by other companies, contributing to the ecology around Max. In 1999, the Netochka Nezvanova collective released a suite of externals that added extensive real-time video control to Max, making the program a great resource for multimedia artists. This kind of live video work had been intuited and worked on by Laurie Spiegel and her work with VAMPIRE, and by Don Slepian who developed his Live Performance Theatrical Visual Instrument in the early 1980s. In 1984, Slepian did a week-long residence as a live video artist at the Pompidou Center, and was shown around IRCAM by Jean-Claude Risset.

The development of Max/MSP got to the point where it was adopted by John Chowning and Max Mathews himself, with the latter learning how to use the program named after him well into his golden years. In 2011, Mathews died of old age and complications from pneumonia. He was 84.

CHANT

Of the many original pieces of software created at IRCAM, the program CHANT, developed by Xavier Rodet, Yves Potard, and Jean-Baptiste Barrière is of special note for its ability to synthesize the singing human voice. Developed between 1979 and 1983, the program was able to model the formants created by the human vocal tract, yet was given enough flexibility for programmers to use it for synthesizing instrumental sounds and noises, such as bells, cymbals, woodwinds and many others. It even gave composers the ability to go beyond traditional vocal and instrumental music altogether. John Chowning was at IRCAM as CHANT was being developed and he lent his musicians ear to the work of the engineers to improve the sound of their software.

CHANT used a synthesis-by-rule model and its creators made subprograms, which can be thought of as presets, to use for specific types of singing, such as bel canto voice for Western style soprano singing and a Tibetan chant mode. In the bel canto subprogram, they used a phase vocoder to analyze the same pitch as interpreted by a number of different singers. With this data they were able to obtain the precise frequencies of the first eight formants used by the singer. For the Tibetan chant subprogram their main concern was to develop a system for voice emulation that accounted for noise and strange harmonics, in contrast to the typical voice of the trained Western singer who tries to eliminate randomness and regional accents.

CHANT was equipped with a number of basic parameters for relative ease of use, but those who sought total compositional control could use an extended version of the program that allowed for different models to be implemented, including those for creating instrumental sounds. CHANT began with analyzing and mimicking vocal behavior, but was capable of going beyond vocal behavior into other areas of sound, including that of granular textures that opened up a variety of possibilities for spectral exploration. With the models in place, composers can work with these definitions to create "imaginary hybrid instruments" that give them and listeners a chance to explore new timbral spaces. Some of these possibilities offered by CHANT have been explored by a number of different composers, including Jonathan Harvey, Jukka Tiensuu, and Tod Machover.

Jonathan Harvey's Ritual Melodies

Born in 1939, Jonathan Harvey was a British composer who liked to jump across the boundaries of genre within contemporary classical music. He had begun his studies with composer Benjamin Britten, who advised him to learn also from Erwin Stein and Hans Keller. Like many other composers in his general age group, he fell under the spell of Karlheinz Stockhausen and attended his composition courses at Darmstadt in 1966 and 1967. In 1969, he got a Harkness Fellowship at Princeton University where he was able to study under Milton Babbitt.

The Balearic Islands of Spain have a history of being good for music, and Harvey wrote his 1973 piece "Inner Light I" while staying in Menorca. It is an electroacoustic work for seven instruments and tape dedicated to Benjamin Britten on the occasion of his sixtieth birthday. Harvey realized the tape portion while sequestered away inside the studios of Swedish Radio in Stockholm and at University College in Cardiff, Wales. This electronic portion features ring modulation and varispeed tape.

Unlike many of his fellow composers on the experimental end of the spectrum, some of Harvey's works are played with frequency, rather than just being concerned with frequency. This is in part due to his early religious affiliations with the Church of England, and his own time as a chorister at St. Michael and All Angels Church in Tenbury. Harvey loved choral music and wrote pieces for the British cathedral choirs. His religious pieces

"I Love the Lord" (1976) and "The Angels" (1992) are thus the most recorded and performed of his music.

Harvey followed the path of many other twentieth century composers and went on to teach composition, working at Southampton and Sussex Universities while doing stints as a guest lecturer in the United States. He was happy to encourage his students and help them develop in their own ways, rather than demanding anyone adhere to a particular school of musical thought. He hadn't, so why should they?

Throughout his career, he would flit between electroacoustic works, purely electronic pieces, and orchestral pieces that utilized live electronics. A number of works he wrote concerned the nature of speech—whether sung, spoken, or synthesized—and its relationship to song. "Mortuous Plango, Vivos Voco" is a short work for eight-channel tape. It uses concrète singing of his son, then a chorister in the Winchester Choir, and the recorded sounds of the largest bell of Winchester Cathedral, transformed in various ways by the use of MUSIC V and CHANT. Other synthesized sounds were also used. The piece also uses phonetics, linguistic analysis, proportions from the Golden Ratio, the judicious use of spatialization, and a sonorous reverb that gels it all together.

The voice of the bell is strong in this work. The title was taken from the Latin words inscribed on the bell, which translate to "I lament the dead, I call the living." The ethereal work is one of genius, recalling the similar use of concrète voices and electronic techniques used in Stockhausen's "Gesang der Junglinge." Like Stockhausen, Harvey was completely open about his mysticism, and his belief in spiritual realities shines through in his music. In spiritual matters he was also as eclectic as he was in his compositions. He had a pronounced interest in Eastern religions, which he seemed to be as comfortable writing music about as he was within the Christian milieu.

Case in point, Harvey's epic "Bhakti" was written in 1982 as a commission from IRCAM and is a piece for fifteen instruments and tape. The structure of the close to hour-long composition is based around texts from the Hindu Rig Veda, which give it a meditative and contemplative aspect. Twelve short movements, each varied three times, give it thirty-six subsections, each of these defined by a certain grouping of instruments playing a particular pitch cell. Showing his serialist leanings, "Bhakti" explores the partials of a single pitch, a quarter-tone above G, below A440. The series are made from proportional intervals above and below that frequency, with space for what Harvey calls "glossing," or allowing for improvisation in devising the pitch cells. The tape part of "Bhakti" was made using sounds from the instrumental ensemble mixed and transformed by the computer. At the end of each movement, a quotation from the Rig Veda is heard, with Harvey considering these 4,000-year-old hymns "keys to consciousness."

Harvey used synthesized voices and instruments again in his 1990 electronic piece "Ritual Melodies." Realized at IRCAM with the help of Jan Vandenheede and the program Formes, which had been designed originally as Computer Assisted Composition environment for CHANT, Vandenheede created a number of sounds using the program. These again included voices, both Western plainchant and Tibetan style chant. The other instrument sounds were all decidedly Eastern, and included a Vietnamese koto, an Indian oboe, Japanese shakuhachi, and a Tibetan bell. Listening to these instruments and voices, they do not sound at all artificial, voice synthesis having come a long way since the days of "Daisy Bell." All of the instruments used are for ritual or religious purposes in different cultures, but Harvey wanted to bring them together in a way that wouldn't normally happen in real world rituals. Here Harvey composed sixteen melodies that seamlessly move between the different synthesized instruments, and form an intertwined circular chain when other melodies are introduced and morph into each other. He writes of the piece that, "Each melody uses the same array of pitches, which is a harmonic series omitting the lowest 5 pitches. Each interval, therefore, is different from every other interval. So the piece as a whole reflects the natural acoustic structure of the instruments and voices." The bell sounds are used to mark different movements of the piece. There is something foreign yet familiar about the vocal sounds, as if the words are from a language being heard for the very first time, giving the work a sense of freshness even after repeated listenings.

Speakings

Jonathan Harvey was as happy to work with traditional instruments and timbres as he was making purely electronic works or purely choral works. He was also happy to mix and match, drawing his influences from a diverse grouping of musicians and teachers. All of these influences are present in his own diversity of work. His ability to work back and forth between modes gave him a lot of freedom, even if it made it hard for critics to pigeonhole his music.

Between 2005 and 2008, Jonathan Harvey was a composer in residence with the BBC Scottish Symphony Orchestra, birthing three major works known together as the *Glasgow Trilogy*. The trilogy begins with "Towards a Pure Land…" (2005), continues with "Body Mandala" (2006), and finishes with the masterpiece "Speakings" (2008). All three pieces combine orchestral instruments with electronics, and all three are inspired by the Buddhist side of his spiritual inclinations, but it is "Speakings" where Harvey once again looks into the correlations between speech and song. Within that same time span Harvey wrote "Sprechgesange" (2007) for oboe, cor anglais, and ensemble, and it is these two pieces that we will look at here.

By using "Sprechgesange" as its title, Harvey ties his purely instrumental piece to the earlier efforts of Schoenberg and Berg. Harvey's idea for "Sprechgesange" came from musing on the psychological roots of speech and sound and how those are so often connected to the cooing voice, talking and singing of the mother to her baby child, who experiences these first in the womb, and then as a newborn and baby, in the very early process of learning to speak and sing themselves. Halfway through the piece, Harvey inserts a musical Wagner reference, calling it "a moment when Parsifal 'hears' the long-forgotten voice of his dead mother call the name, his own name, that he had forgotten - an action of the shamanistic Kundry. From this awakening, this healing, comes the birth of song from the meaningless chatter of endless human discourse." Again referencing the title, he adds, "'Speech' with deep meaning..."

"Speakings" was commissioned in part by IRCAM and Radio France, who helped with the electronic side of things, again using programs to synthesize speech. Harvey uses the orchestral palette to make further voicings that mimic the utterance of phonemes, building on the techniques he had used in "Sprechgesange." From the slow beginning, the organic and the digital merge together into a gradual towering babble of enunciation by the second movement. The tracery of vocoded signals is laced into the chaos of linguistic polyphony, with Harvey writing of the piece:

> *The orchestral discourse, itself inflected by speech structures, is electro-acoustically shaped by the envelopes of speech taken*

from largely random recordings. The vowel and consonant spectra-shapes flicker in the rapid rhythms and colours of speech across the orchestral textures. A process of 'shape vocoding', taking advantage of speech's fascinating complexities, is the main idea of this work.

Different instruments had the "shape vocoding" applied to them through the judicious use of microphones.

The third and final movement begins with bell rings and horn blasts, weaving between each other in a way that is reminiscent of how mantras are intoned with full vibration. The listener is now in a sacred place, a cathedral or temple, and the voices here chant an incantatory song, along single monodic lines that reverberate through space. Here we return to what Harvey says is the "womb of all speech," the Buddhist mantra OM-AH-HUM, which in the mythology of India is said to be half-song, half-speech. This is pure speech. The original tongue. In Judaeo-Christian terms, it can be likened to the original language spoken by Adam and Eve before the time of Babel, before the time when humanities primordial tongue was shattered and split into multiplicity.

A FINAL WORD

In the beginning was the word. It was tapped out in morse code tones, sent over telegraph wires, and spoken into microphones to travel down telephone lines, or go out over antennas into radio waves to skip across the aether. Electricity and recording technology made sound malleable, and the instruments born in the telecommunications laboratories were wing footed and quick to escape into the hands of enterprising musicians. The point has long since passed when the engineer and musician joined forces to create the music engineer.

The dreams of the telecommunications pioneers have been achieved. Synthesized music is piped in over the phone lines, much as Thaddeus Cahill thought it might, providing soothing music for insomniacs, background ambience for those at work, and the sense of emotional resonance and human connection people listen to music for in the first place. The spread spectrum technology of Hedy Lammar and George Antheil is now used every time someone logs onto a WiFi network. The children of the vocoders developed by Home Dudley get used whenever someone picks up a cellphone to call their friends and loved ones, or when a rapper shunts their voice through Auto-Tune.

At the same time, the interlaced worlds of classical art music and academic composers are not as big on the cultural stage as they once were. The computer works that on first arrival sounded novel, are no longer novel. We have become accustomed to the new electronic timbres. The common practice of tonal harmony and general principles that once bound composers and musicians together has now been splintered, in the same way the atom was splintered, in the same way the word was broken down into its constituent parts for later reassembly. We now have what composer Alvin Curran calls a "new common practice," a grab bag of tools and techniques that have evolved over the twentieth century to the present. These can now be recombined and explored at the total whim of each individual musician, composer, band, or project as they follow their muse.

The tools birthed by avant-garde composers and engineers, in the early studios supported by radio stations and research centers, have now multiplied and proliferated. Powerful programs and gear, once only available in places like Bell Labs, leaked out into the recording studios, and are now in the attics, basements, and spare rooms of home producers and music lovers. The very nature of sound, acoustics and perception are now able to be explored in these alchemical laboratories where new music can be fermented, distilled and prepared for listeners to imbibe. Some of these creations go on to become the slabs of vinyl wax and discs of shiny plastic bought in the record stores and streamed over the internet.

Home recording reinvented music. With the means of production in the hands of the musicians, it was no longer up to the major labels to decide what kind of music to record and dish out to the public. A variety of musical styles simmered in the underground as the equipment to make electronic music reached the music makers, the dreamers of dreams. Much of it ended up on the sound systems of clubs and at raves, music for the dancers; others used these tools to craft ambient soundtracks for the intrepid psychonauts tripping in chill out rooms and bedrooms, or just for regular listeners eager to relax. Extreme forms of noise and experimental head-scratchers also trickled out of the scrappy studios built by those with a passion for avant-audio. The methods and means have made their way into all kinds of popular music, niche genres, and many modes of multimedia production.

At the same time, as sound engineering tools became democratized, there was continued interest in technology and business to develop and refine speech synthesis and create machines capable of conversation. It is an industry now worth tens of billions of dollars. Speech synthesis has come such a long way, to the point where the voices of famous singers and actors can now be cloned so well that the average listener finds it hard to detect the discrepancies between actual recordings of the human singer and a synthesized digital simulacra of a singer. This has staggering implications for entertainment, as well as our fragile media ecosystems. In a world where it is already difficult to tell what news is real and what is propaganda, the ability for anyone with the right software to deepfake someone's voice and line it up with a video of them saying those words has the ability to wreak all kinds of havoc.

The same telecommunications companies and researchers that worked so hard to develop efficient means for connecting the world are now capable of spreading this language virus far and wide. The online video that someone thinks is from a trusted public figure may actually have been made by a disenfranchised kid somewhere in flyover country, the forgotten states of America. The sound of a politician from one country can now be easily mimicked to say something embarrassing or provocative and put out on social media by foreign agents with bad intentions for seeding political divisions. It's not just the threat of agitated political chaos that is problematic, but criminals are now using voice synthesis software to mine social media and target friends and families of people, faking the voices of their loved ones in the service of scams and frauds. As with many sharp tools, speech synthesis is double edged.

As contentious as sampling the work of other artists has been over the years, the availability of software that can mimic the voices of Dolly Parton, Bob Dylan, or Bruce Springsteen, and have them sing things they wouldn't write, will continue to raise issues about artistic creation in what theorist Walter Benjamin foresaw as the "age of mechanical reproduction." The voices of dead musicians can now be resurrected and put to work in supergroups who otherwise may never have been, to usher in a new era of automatic music for the people. For those of us who delight in surrealistic sound collages, mashups, and remix culture, the use of voice synthesis could create a fascinating next chapter for electronic music.

The amount of intelligence and insight that scientists, musicians and researchers put into exploring the nature of speech, acoustics, and perception was massive. Powerful technology requires an equal mentality for its creative use. Speech and music synthesis might best be put in service to the human imagination as tools for exploring the inner essence of sound itself. Now our own sense of discernment must be honed if it is not the machines themselves who are to have the final word.

SELECTED BIBLIOGRAPHY

FIRST UTTERANCE

Bacon, Francis. New Atlantis. Los Angeles, CA.: The Philosophical Research Society, 1985.

Jackson, Myles W. "Harris Lecture 5-5-2022" YouTube, May 11, 2022. 10:45, www.youtube.com/watch?v=dmCpmJOCF-w

PART I: TELEMUSIK

1: TELEGRAPHIES

VISIBLE SPEECH

Bruce, Robert V. *Bell: Alexander Bell and the Conquest of Solitude*. Ithaca, New York.: Cornell University Press, 1990.

Duchan, Judith Felson. *The Phonetic Notation System of Melville Bell and its Role in the History of Phonetics*. The Journal of Speech-Language Pathology and Audiology Vol. 30. No. 1, Spring 2006

Denis Larouche, Maker. *Ear Phonautograph (Reconstruction)*. https://soundandscience.de/instrument/ear-phonautograph-reconstruction

Novak, David, Sakakeeny, Matt, ed. *Keywords in Sound*. Durham, North Carolina.: Duke University Press, 2015.

THE MUSICAL TELEGRAPH & TELHARMONIOUM

Coe, Lewis. *The Telephone and its Several Inventors: a History.* Jefferson, NC.: McFarland & Co. 1995

Crab, Simon. "The 'Musical Telegraph' or 'Electro-Harmonic Telegraph', Elisha Gray. USA, 1874." Crab, http://120years.net/the-musical-telegraphelisha-greyusa1876

Crab, Simon. "The 'Telharmonium' or 'Dynamophone' Thaddeus Cahill, USA 1897."http://120years.net/the-telharmonium-thaddeus-cahill-usa-1897/

Dewan, Brian. "Thaddeus Cahill's 'Music Plant.'" https://www.cabinet-magazine.org/issues/9/dewan.php

Lapsley, Phil. Exploding the Phone: The Untold Story of the Teenagers and Outlaws who Hacked Ma Bell. New York, NY.: Grove Press, 2013.

Martin, Thomas Commerford. "The Telharmonium: Electricity's Alliance With Music." http://earlyradiohistory.us/1906telh.htm

Peck, William. Commercial Broadcasting Pioneer: The WEAF Experiment 1922-1926. Cambridge, MA.: Harvard University Press, 1946.

Shulman, Seth. *The Telephone Gambit: Chasing Alexander Graham Bell's Secret.* New York, NY.: W.W. Norton & Co., 2008.

Standage, Tom. *The Victorian Internet: the Remarkable Story of the Telegraph and the Nineteenth Century's On-line Pioneers.* New York, NY.: Walker Publishing Company, 1998.

Weidenaar, Reynold. Magic Music from the Telharmonium. Metuchen, N.J.: Scarecrow Press, 1995.

White, Thomas H. "Electric Telephone" http://earlyradiohistory.us/sec003.htm

Williston, Jay. "Thaddeus Cahill's Teleharmonium" http://www.synthmuseum.com/magazine/0102jw.html

2: WIRELESS FANTASIES

THE SINGING ARC

Crab, Simon. The 'Singing Arc', William Duddell, UK, 1899 http://120years.net/the-singing-arcwilliam-duddeluk1899/

Engineering and Technology History Wiki. "Poulsen-Arc Radio Transmitter, 1902" https://ethw.org/Milestones:Poulsen-Arc_Radio_Transmitter,_1902

Hong, Sungook. Wireless: From Marconi's Black Box to Audion. Cambridge, MA.: MIT Press, 2001.

Sussman, Herbert. *Victorian Technology: Invention, Innovation, and the Rise of the Machine*. Santa Barbara, CA.: ABC-CLIO 2009

Today in Science. "William Du Bois Duddell" https://todayinsci.com/D/Duddell_William/DuddellWilliamBio.htm

United States Army Signal Corps. *Principles Underlying Radio Communications.* Washington, D.C. National Bureau of Standards, 1922.

THE AUDIO PIANO

Archer, Gleason L. *History of Radio to 1926*. New York, NY.: The American Historical Society, 1938

Crab, Simon. "The 'Audion Piano' and Audio Oscillator. Lee De Forest. USA, 1915." https://120years.net/the-audion-pianolee-de-fores-tusa1915/

De Forest, Lee. The Father of Radio. Chicago, IL.: Wilcox & Follet Co., 1950.

Harlow, Alvin F. Old Wires and New Waves: The History of the Telegraph, Telephone and Wireless.:

NY, New York.: D. Appleton-Century Company, 1936.

Sungook, Hong. Wireless: From Marconi's Black Box to Audion. Cambridge, MA.: MIT Press, 2001

3: VIBRATIONS FROM THE AETHER

LEV THEREMIN AND THE VIBRATIONS OF THE AETHER

Crypto Museum. "The Thing: The Great Seal Bug." https://www.crypto-museum.com/covert/bugs/thing/index.htm

Glinsky, Albert. Theremin: Ether Music and Espionage. Chicago, IL.: University of Illinois Press, 2000

TWO PIONEERS OF SPREAD SPECTRUM RADIO

Antheil, George. *Bad Boy of Music.* Garden City, N.Y.: Doubleday, 1945.

Barton, Ruth. *Hedy Lamarr: The Most Beautiful Woman in Film.* Lexington, KY.: University of Kentucky Press, 2010.

Historical Notes: The Fantastic Lives of Hedy Lamarr. https://ima.org.uk/16506/historical-notes-the-fantastic-lives-of-hedy-lamarr/

Kahn, David. *How I Discovered WWII's Greatest Spy and Other Stories of Intelligence and Code.* Boca Raton, FL.: CRC Press, 2014

Lamarr, Hedy. Ecstasy and Me: My Life as a Woman. London, UK.: W.H. Allen, 1967.

Rhodes, Richard. *Hedy's Folly: The Life and Breakthrough Inventions of Hedy Lammar, the Most Beautiful Woman in the World.* New York, NY.: Vintage, 2012.

Richter, Hans. *Dada: Art and Anti-Art.* Thames and Hudson, 1965.

Sitsky, Larry, ed. *Music of the 20th Century Avant-Garde: A Biocritical Sourcebook.* Westport, CT.: Greenwood Press, 2002.

PART II: THE VOICE OF THE BELL

4: ENCIPHERED SOUNDS

AUDITORY PERCEPTION AND ARTICULATION

Brigham Young High School. *Harvey Fletcher: Scientist, Father of Stereophonic Sound, Author.*

http://www.byhigh.org/History/Fletcher/DrHarvey.html

Fletcher, Stephen H. *Harvey Fletcher, 1884—1981, A Biographical Memoir.* http://www.nasonline.org/publications/biographical-memoirs/memoir-pdfs/fletcher-harvey.pdf

Washington, D.C.: National Academy of Sciences, 1992.

THE VOCODER AND THE VODER

Dudley, Homer. "The Carrier Nature of Speech." The Bell System Technical Journal, Vol. 19, No. 4, October 1940

Dudley, Homer. "Fundamentals of Speech Synthesis." Journal of the Audio Engineering Society, Vol. 3, No. 4, October 1955

Mills, Mara. "Media and Prosthesis: The Vocoder, the Artificial Larynx, and the History of Signal Processing." Qui Parle: Critical Humanities and Social Sciences, vol. 21 no. 1, 2012

Tompkins, Dave. *How to Wreck a Nice Beach: The Vocoder from WWII to Hip-Hop: The Machine Speaks*. Brooklyn, NY.: Melville House, 2010.

Wyndam, Ackley. "The World's First Talking Machine." https://liwaiwai.com/2019/05/27/the-worlds-first-talking-machine/

SIGSALY: CRYPTOGRAPHY, TURNTABLES AND MUZAK

Boone, J.V., Peterson, R.R. "SIGSALY: The Start of the Digital Revolution." https://webharvest.gov/peth04/20041022071732/http://www.nsa.gov/publications/publi00019.cfm

Christensen, Chris. "SIGSALY" https://www.nku.edu/~christensen/SIGSALY.pdf

Kahn, David. How I Discovered WWII's Greatest Spy and Other Stories of Intelligence and Code.

Boca Raton, FL.: CRC Press, 2014

Tompkins, Dave. *How to Wreck a Nice Beach: The Vocoder from WWII to Hip-Hop: The Machine Speaks*. Brooklyn, NY.: Melville House, 2010.

5: MUSIC BY NUMBER: THE MATHEMATIKOI

AUDREY: SPEECH RECOGNITION

Moskvitch, Katia. "The Machines that Learned to Listen." https://www.bbc.com/future/article/20170214-the-machines-that-learned-to-listen

Juang, B.H., Rabiner, Lawrence R. "Automatic Speech Recognition – A Brief History of the Technology Development." Georgia Institute of Technology, Rutgers University and the University of California.

Warren, Jennifer. Audrey: the First Speech Recognition System. https://astaspeaks.wordpress.com/2014/10/13/audrey-the-first-speech-recognition-system/

TAKING IT TO THE MAX: MUSIC I-III

Chasalow, Eric. "Max Matthews on MUSIC I - the beginning of computer music - Bell Labs." https://www.youtube.com/watch?v=mT3U98cFqSs

Dayal, Geeta. *Max Mathews (1926-2011).* https://web.archive.org/web/20130115160847/http://blog.frieze.com/max-mathews/

Di Nunzio, Alex. "Music I." https://www.musicainformatica.org/topics/music-11-2.php

Dubois, R. Luk. "The First Computer Musician." https://archive.nytimes.com/opinionator.blogs.nytimes.com/2011/06/08/the-first-computer-musician/

Holmes, Thom. *Electronic and Experimental Music: Technology, Music, and Culture.* New York, NY.: Routledge, 2020.

Manning, Peter. *Electronic and Computer Music.* Oxford, UK.: Clarendon Press, 1993.

A BICYCLE BUILT FOR TWO

IBM 7090 Computer to Digital Sound Transducer. *Music from Mathematics.* Decca LP 9103. 1962.

Haas, Jeffrey. "Follow-on Vocoder and Speech Synthesis Technology" in Introduction to Computer Music, online book. https://cmtext.indiana.edu/synthesis/chapter4_speech_vocoder6.php

The New York Public Library, Archives & Manuscripts. "Irving Teibel, Biographical and Historical Information" https://archives.nypl.org/rha/24592#bioghist

Pierce, John R. *Science, art, and communication.* New York, NY.: C.N. Potter, 1968.

Poundstone, William. John Kelly. http://home.williampoundstone.net/Kelly.htm

Smith, J.O. "Singing Kelly-Lochbaum Vocal Tract", in *Physical Audio Signal Processing*, online book https://ccrma.stanford.edu/~jos/pasp/Singing_Kelly_Lochbaum_Vocal_Tract.html

Teibel, Irv. "Environments 1" https://www.irvteibel.com/discography/environments/

MUSIC FROM MATHEMATICS

Melville Jr., Clark. "Science and Technology in the Fine Arts: Music from Mathematics." Science, 4 Jan 1963, Vol 139, Issue 3549, pp. 28-29

IBM 7090 Computer to Digital Sound Transducer. *Music from Mathematics*. Decca LP 9103. 1962.

Don Slepian, e-mail message to author, December 8, 2023.

MUSIC IV-V

Buffalo News. "Louis J. Gertsman Dies; Research Produced First Talking Computer." https://buffalonews.com/news/louis-j-gerstman-dies-research-produced-first-talking-computer/article_80459d6a-c0d3-5412-ba71-0039264d1e48.html

Haas, Jeffrey. "Introduction to Computer Music: An Electronic Textbook." https://cmtext.indiana.edu/index.php

Holmes, Thom. *Electronic and Experimental Music: Technology, Music, and Culture*. New York, NY.: Routledge, 2020.

Manning, Peter. *Electronic and Computer Music*. Oxford, UK.: Clarendon Press, 1993.

The Wire. "French Electronic Music Composer Jean-Claude Risset Has Died." https://www.thewire.co.uk/news/44612/french-electronic-music-composer-jean-claude-risset-has-died

Ircam. "Jean-Claude Risset" https://brahms.ircam.fr/en/jean-claude-risset#bio

Risset, Jean-Claude. "Max Mathews's Influence on (My) Music." Computer Music Journal, Vol. 33, No. 3 (Fall, 2009), pp. 26-34.

6: VARIATIONS FOR SPEECH

ROBERT MOOG, WENDY CARLOS AND A CLOCKWORK ORANGE

Holmes, Thom. *Electronic and Experimental Music: Technology, Music, and Culture*. New York, NY.: Routledge, 2020.

Pinch, Trevor, Trocco, Frank. *Analog Days: The Invention and Impact of the Moog Synthesizer*. Cambridge, MA.: Harvard University Press, 2002.

Holmes, Thom (2008). Electronic and Experimental Music: Technology, Music, and Culture

Prochnow, David. *Talk: Projects in Speech Synthesis*. New York, NY.: Tab Books, 1987.

Warpcore. "RIP Florian Schneider: How Kraftwerk became the forefathers of modern electronic music." https://warpcore.medium.com/rip-florian-schneider-how-kraftwerk-became-the-forefathers-of-modern-electronic-music-b47740d814a9

LINEAR PREDICTIVE CODING, FUMITADA ITAKURA, MANFRED SCHROEDER AND BISHNU S. ATAL

Digital Mobile Radio Association. "Electromagnetic compatibility and Radio spectrum Matters (ERM); Digital Mobile Radio (DMR) General System Design" https://www.dmrassociation.org/downloads/standards/tr_102398v010401p.pdf

Engineering and Technology History Wiki "Bishnu S. Atal" https://ethw.org/Bishnu_S._Atal

Lugosch, Loren. "Predictive Coding in Machines and Brains." https://lorenlugosch.github.io/posts/2020/07/predictive-coding/

Nebeker, Frederik. "Fumitada Itakura: an Oral History" https://ethw.org/Oral-History:Fumitada_Itakura

Nebeker, Frederik. "Manfred Schroeder: an Oral History" https://ethw.org/Oral-History:Manfred_Schroeder

Rowetel. "Codec 2" http://www.rowetel.com/?page_id=452/

GODFREY WINHAM, KENNETH STEIGLITZ, PAUL LANSKY

Paul Lansky, composer, in discussion with the author, December 15, 2023.

"Idle Chatter by Paul Lansky" https://music7703lsu.wordpress.com/2017/04/29/idel-chatter-by-paul-lanky/

Lansky, Paul. "More Than Idle Chatter" ttp://paul.mycpanel.princeton.edu/liner_notes/morethanidlechatter.html

"Paul Lansky – Mild Und Leise" https://www.tumblr.com/post-punk/149013720/paul-lansky-mild-und-leise-on-idioteque-the

Lansky, Paul. "Reflections on Spent Time" https://paul.mycpanel.princeton.edu/lansky-icmc-keynote.pdf

Nathans, Aaron. "Composers & Computers" https://engineering.princeton.edu/news/2022/05/05/episode-1-serialism

https://engineering.princeton.edu/news/2022/05/12/episode-3-converter

https://engineering.princeton.edu/news/2022/05/19/episode-4-idle-chatter

Ranta, Alan. "Idle Chatter About Paul Lansky" https://www.popmatters.com/141865-idle-chatter-about-paul-lanskys-notjustmoreidlechatter-2496021210.html

SPEECH SONGS

Hougland, Eli. "Charles Dodge- Earth's Magnetic Field & Bell Laboratories" https://info.umkc.edu/specialcollections/archives/1056

Nathans, Aaron. "Composers & Computers" https://engineering.princeton.edu/news/2022/05/19/episode-4-idle-chatter

Perfect Sound Forever. "Charles Dodge on 'Speech Songs'." https://www.furious.com/perfect/ohm/dodge.html

SPEAK AND SPELLING WITH Q. REED GHAZALA

Ghazala, Q. Reed. *Circuit-Bending: Build Your Own Alien Instruments.* Indianapolis, IN.: Wiley, 2005.

Ghazala, Q. Reed. http://www.anti-theory.com/

Gross, Jason. "Reed Ghazala" http://www.furious.com/perfect/emi/reedghazala.html

Kirn, Peter. "The Strange Cartridge Powered Speech of TI Touch & Tell" https://cdm.link/2018/07/the-strange-cartridge-powered-speech-of-ti-touch-tell/

Various Artists. Gravikords, Whirlies and Pyrophones: Experimental Musical Instruments, written and produced by Bart Hopkins. Ellipsis Arts, 1998.

PART III. WE ALSO HAVE SOUND-HOUSES

7 HEARING INNER SOUNDS: THE AKOUSMATIKOI

ARTS OF NOISE

Cox, Christoph, Warner, Daniel, ed. *Audio Culture: Readings in Modern Music.* New York, NY .:Bloomsbury Academic, 2017

Wen-Chung, Chou. "Open Rather Than Bounded" https://chouwen-chung.org/writing/excerpts-from-open-rather-than-bounded/

EXPRESSIONS OF ZAAR

Scene Arabia. "The Father of Electronic Music: A Brief History of Egyptian Composer Halim El Dabh" https://scenearabia.com/Noise/halim-el-dabh-egyptian-musician-composer-invented-electronic-music

Seachrist, Denise. The Musical World of Halim El Dabh. Kent, OH.: Kent State University Press, 2003.

Pierre Schaeffer: Musician of Sounds, Club D'Essai and Inside the Acousmatic Soundspace

Battier, Marc. "What the GRM Brought to Music: From Musique Concrète to Acousmatic Music." https://music.arts.uci.edu/dobrian/CMC2009/OS12.3.Battier.pdf

Bayle, Francois. "Biography" http://www.francoisbayle.fr/?page_id=1214

Crab, Simon. "The 'Groupe de Recherches Musicales' Pierre Schaeffer, Pierre Henry & Jacques Poullin, France 1951" https://120years.net/wordpress/the-grm-group-and-rtf-electronic-music-studio-pierre-schaeffer-jacques-poullin-france-1951/

Holmes, Thom. *Electronic and Experimental Music: Technology, Music, and Culture.* New York, NY.: Routledge, 2020.

INA grm - Groupe de Recherches Musicales. "Machines, Tools, Studios" https://artsandculture.google.com/story/YAVhtqcihjlQIw

Sitsky, Larry, ed. Music of the 20th Century Avant-Garde: A Biocritical Sourcebook. Westport, CT.: Greenwood Press, 2002.

Wishart, Trevor. *On Sonic Art.* Amsterdam, Netherlands.: Harwood Academic Publishers, 1996.

Stockhausen, Karlheinz. 1992. "Etude (1952): Musique Concrète". Booklet for Karlheinz Stockhausen: Elektronische Musik, 5–7 (German) and 95–100 (English). Stockhausen Complete Edition CD 3. Kürten: Stockhausen-Verlag.

8: ELECTRIC OSCILLATIONS: THE STUDIO FOR ELECTRONIC MUSIC OF THE WEST GERMAN RADIO

DR. FRIEDRICH TRAUTWEIN AND THE RADIO EXPERIMENTAL LABORATORY

Birdsall, Carolyn. "Radio Documents: Broadcasting, Sound Archiving, and the Rise of Radio Studies in Interwar Germany." Technology and Culture 60, no. 2 (2019): S96-S128.

Draper, Charlie. "Oskar Sala Plays Genzmer's Trautonium Concerto No. 1" https://charliedraper.com/articles/2018/12/13/oskar-sa-la-plays-genzmers-trautonium-concerto-no-1

Jackson, Myles W. "German Radio and the Development of Electric Music in the 1920s and 1930s." https://www.mpiwg-berlin.mpg.de/research/projects/german-radio-and-development-electric-music-1920s-and-1930s

Jackson, Myles W. "Harris Lecture 5-5-2022" YouTube, May 11, 2022. https://www.youtube.com/watch?v=dmCpmJOCF-w

HARALD BODE AND THE HEINRICH HERTZ INSTITUTE FOR RESEARCH ON OSCILLATIONS

Birdsall, Carolyn. "Radio Documents: Broadcasting, Sound Archiving, and the Rise of Radio Studies in Interwar Germany." Technology and Culture 60, no. 2 (2019): S96-S128.

Crab, Simon. "The 'Melochord', Harald Bode, Germany, 1947" https://120years.net/wordpress/the-melochordharald-bodegermany1947/

Palov, Rebekkah. "Harald Bode — A Short Biography" https://econtact.ca/13_4/palov_bode_biography.html

GENESIS OF THE STUDIO FOR ELECTRONIC MUSIC WDR

Chang, Ed. "Stockhausen on Electronic Music (1952-1960)" and "WDR Electronic Music Studio Tour (2015)" https://stockhausenspace.blogspot.com/2015/08/stockhausen-on-electronic-music-wdr.html

Crab, Simon. "WDR Electronic Music Studio, Werner Meyer-Eppler, Robert Beyer & Herbert Eimert, Germany, 1951" https://120years.net/wordpress/wdr-electronic-music-studio-germany-1951/

Holmes, Thom. *Electronic and Experimental Music: Technology, Music, and Culture.* New York, NY.: Routledge, 2020.

Maconie, Robin. *Other Planets: The Complete Works of Karlheinz Stockhausen 1950-2007.* Lanham, MA.: Rowman & Littlefield, 2016.

STUDIES IN ELECTRONICS

Chang, Ed. "Etude, Studie I & II" https://stockhausenspace.blogspot.com/2014/12/opus-3-studie-i-studie-ii-and-etude.html

Kryzaniak, Mike. "Stockhausen's Studies I and II" https://michaelkrzyzaniak.com/Research/Stockhausen_Studie_II/Maconie, Robin. *Other Planets: The Complete Works of Karlheinz Stockhausen 1950-2007.* Lanham, MA.: Rowman & Littlefield, 2016.

Nordin, Ingvar Loco. "Stockhausen Edition no. 3 (Electronic Music 1952 – 1960)" https://www.sonoloco.com/rev/stockhausen/03.html

GESANG DE JUNGLINGE

Maconie, Robin. *Other Planets: The Complete Works of Karlheinz Stockhausen 1950-2007.*

Lanham, MA.: Rowman & Littlefield, 2016.

Langer, Ken. "The Ambient Church: Seeking the Spiritual Through the Power of Music."https://sites.google.com/site/klangerdude/home/ministry/papers/the-ambient-church-movement

Smalley, John. "Gesang der Jünglinge: History and Analysis" http://sites.music.columbia.edu/masterpieces/notes/stockhausen/GesangHistoryandAnalysis.pdf

MAKING TELEMUSIK IN JAPAN

Dayal, Geeta. "Stockhausen in Japan." https://daily.redbullmusicacademy.com/2014/10/stockhausen-in-japan

Holmes, Thom. *Electronic and Experimental Music: Technology, Music, and Culture.* New York, NY.: Routledge, 2020.

Nordin, Ingvar Loco. "Stockhausen Edition no. 9 (Mikrophonie I & II / Telemusik)" https://www.sonoloco.com/rev/stockhausen/stockhausen.html

Shibata, Minao. "Music and Technology in Japan." https://www.ubu.com/media/text/emr/books/UNESCO/12_Shibata.pdf

HYMNEN

Holmes, Thom. *Electronic and Experimental Music: Technology, Music, and Culture.* New York, NY.: Routledge, 2020.

Maconie, Robin. *Other Planets: The Complete Works of Karlheinz Stockhausen 1950-2007.*

Nordin, Ingvar Loco. "Stockhausen Edition no. 47." https://www.sonoloco.com/rev/stockhausen/stockhausen.html

ARTIKULATIONS

Holmes, Thom. *Electronic and Experimental Music: Technology, Music, and Culture.* New York, NY.: Routledge, 2020.

Sitsky, Larry, ed. Music of the 20th Century Avant-Garde: A Biocritical Sourcebook. Westport, CT.: Greenwood Press, 2002.

9: SONIC CONTOURS: COLUMBIA-PRINCETON ELECTRONIC MUSIC CENTER

OTTO LUENING AND VLADIMIR USSACHEVSKY

Columbia-Princeton Electronic Music Center. *10th Anniversary*. New World Records, NWCRL268, 1971.

Computer Music Center. "The Computer Music Center at Columbia University" https://cmc.music.columbia.edu/about

Holmes, Thom. *Electronic and Experimental Music: Technology, Music, and Culture*. New York, NY.: Routledge, 2020.

Shields, Alice. "Vladimir Ussachevsky." https://ubu.com/sound/ussachevsky.html

Sitsky, Larry, ed. *Music of the 20th Century Avant-Garde: A Biocritical Sourcebook*. Westport, CT.: Greenwood Press, 2002.

THE MICROPHONICS OF HARRY F. OLSON, THE RCA MARK I SYNTHESIZER

Crab, Simon. "The 'RCA Synthesiser I & II' Harry Olson & Herbert Belar, USA, 1951." https://120years.net/wordpress/the-rca-synthesiser-i-iiharry-olsen-hebert-belarusa1952/

Harris, Cyril M. *Harry F. Olson 1901—1982: A Biographical Memoir.* Washington, D.C.: National Academy of Sciences, 1989.Holmes, Thom. *Electronic and Experimental Music: Technology, Music, and Culture*. New York, NY.: Routledge, 2020.

Miles, Walter R. *Carl Emil Seashore 1866-1949: A Biographical Memoir.* Washington, D.C.: National Academy of Sciences, 1956.

Osborne, Luka. Remembering the Grateful Dead's Wall of Sound: an Absurd Feat of Technological Engineering. https://happymag.tv/grateful-dead-wall-of-sound/

Snac Cooperative. *Herbert Belar 1901-1997*. https://snaccooperative.org/ark:/99166/w6737t86

MILTON BABBITT: THE MUSICAL MATHEMATICIAN

Babbitt, Milton. *Words About Music*. University of Wisconsin Press, 1987

Duffie, Bruce. "Composers Milton Babbitt." http://www.bruceduffie.com/babbitt.html

Little, Thomas. "Milton Babbitt's Musical Tetris." https://www.youtube.com/watch?v=c9WvSCrOLY4

Little, Thomas. "Set Theory: An Introduction" https://www.youtube.com/watch?v=6BfQtAAatq4

Sitsky, Larry, ed. *Music of the 20th Century Avant-Garde: A Biocritical Sourcebook*. Westport, CT.: Greenwood Press, 2002.

Whittall, Arnold. *The Cambridge introduction to serialism*. Cambridge, UK.: Cambridge University Press, 2008.

VICTOR, WIRED FOR WIRELESS

Cook, Amanda. "Milton Babbitt: Synthesized Music Pioneer." https://betweentheledgerlines.wordpress.com/2013/06/08/milton-babbitt-synthesized-music-pioneer/

Holmes, Thom. *Electronic and Experimental Music: Technology, Music, and Culture*. New York, NY.: Routledge, 2020.

Sitsky, Larry, ed. *Music of the 20th Century Avant-Garde: A Biocritical Sourcebook*. Westport, CT.: Greenwood Press, 2002.

Vladimir Ussachevsky. *Electronic and Acoustic Works* 1957-1972. New World Records, 2007.

MODULATION IN THE KEY OF BODE

Bode, Harlad. "Sound Synthesizer Creates New Musical Effects." http://cec.sonus.ca/econtact/13_4/bode_synthesizer.html

Palov, Rebekkah. "Harald Bode — A Short Biography" https://econtact.ca/13_4/palov_bode_biography.html

10: A NEW ATLANTIS

SOUND HOUSES OF DAPHNE ORAM

Daphne Oram Trust. "Daphne Oram: A Brief Biography" https://www.daphneoram.org/aboutoram/

Houghtaling, Ted. "Edgard Varese – Poeme Electronique" https://www.wnyc.org/story/edgard-varese-poeme-electronique/

Marshall, Steve. "Graham Wrench: The Story Of Daphne Oram's Optical Synthesizer" https://www.soundonsound.com/people/graham-wrench-story-daphne-orams-optical-synthesizer

Oram, Daphne. *An Individual Note of Music Sound and Electronics*. London, UK.: Galliard, 1972.

Oram, Daphne. *Electronic Sound Patterns*. UK. His Master's Voice, 1962.

Patteson, Thomas, Loughridge, Deidre. "Cat Pianos, Sound-Houses, and Other Imaginary Musical Instruments." https://publicdomainreview.org/essay/cat-pianos-sound-houses-and-other-imaginary-musical-instruments
Reynolds, Simon. "What's Behind the Reissue Boom in 'Outsider Electronics'?" https://www.frieze.com/article/music-15
Williams, Holly. "The Woman Who Could Draw Music." https://www.bbc.com/culture/article/20170522-daphne-oram-pioneered-electronic-music
Worby, Robert. "Daphne Oram: Portrait of an Electronic Music Pioneer." https://www.theguardian.com/music/2008/aug/01/daphne.oram.remembered

SPHERICAL VORTICES OF DELIA DERBYSHIRE

Briscoe, Desmond. *The BBC Radiophonic Workshop: The First 25 Years*. UK.: British Broadcasting Corp, 1983.
Butler, David. "Delia Derbyshire." https://www.bbc.com/historyofthebbc/100-voices/pioneering-women/women-of-the-workshop/delia-derbyshire
Cavanage, John. "On Our Wavelength." http://www.delia-derbyshire.org/interview_boa.php
Neibur, Louis. *Special Sound: The Creation and Legacy of the BBC Radiophonic Workshop*. New York, NY.: Oxford University Press, 2010.
Sonic Boom. "Interview with Delia Derbyshire" http://www.delia-derbyshire.org/interview_surface.php
White, Paul. "David Vorhaus: Electronic Music Pioneer."https://www.soundonsound.com/people/david-vorhaus
Wikidelia. https://wikidelia.net/

11: ECHOES OF THE BELL

MAX MATHEWS LIKES TO GROOVE

Dayal, Geeta. *Max Mathews (1926-2011)*. https://web.archive.org/web/20130115160847/http://blog.frieze.com/max-mathews/
Holmes, Thom. *Electronic and Experimental Music: Technology, Music, and Culture*. New York, NY.: Routledge, 2020.

Manning, Peter. *Electronic and Computer Music.* Oxford, UK.: Clarendon Press, 1993.

GROOVING WITH LAURIE SPIEGEL, HARMONICIES MUNDI, AN EXPANDING UNIVERSE

Jet Propulsion Laboratory. "The Golden Record." https://voyager.jpl.nasa.gov/golden-record/

Spiegel, Laurie. "Expanding Universe" http://retiary.org/ls/expanding_universe/index.html

Walls, Seth Colter. "An Electronic Music Classic Reborn." https://www.newyorker.com/culture/culture-desk/an-electronic-music-classic-reborn

Reynolds, Simon. "Resident Visitor: Laurie Spiegel's Machine Music." https://pitchfork.com/features/article/9002-laurie-spiegel/

Lefford, Nyssim and Scheirer, Eric D. Scheirer, and Vercoe, Barry L. "An Interview with Barry Vercoe." https://www.media.mit.edu/events/EMS/bv-interview.html

ALICE IN DIGITAL WONDERLAND

Chabade, Joel. *Electric Sound: The Past and Promise of Electronic Music.* Upper Saddle River, NJ.: Simon & Schuster, 1997.

Crab, Simon. "Bell Labs Hal Alles Synthesiser, Hall Alles, USA, 1976." https://120years.net/wordpress/bell-labs-hal-alles-synthesiser-hall-alles-usa-1977/

Hideaway Studio. "Hideaway Studio Proudly Presents: Synergenesis." https://hideawaystudio.net/2014/09/26/hideaway-studio-proudly-presents-synergenesis/

THE DIGITAL BLISS OF DON SLEPIAN

Don Slepian, e-mail message to author, December 8, 2023.

Slepian, Don. "Electronic Music Pioneer." https://donslepian.com/

Slepian, Don. "Max Mathews and Me." https://synthandsoftware.com/2019/10/max-mathews-and-me/

FROM ALICE TO AMY

Atari. "Amy 1 Spec." http://www.digitpress.com/library/techdocs/AMY_1_spec_confidential_binder_ver_2.pdf

Atarimax. "AMY." https://www.atarimax.com/jindroush.atari.org/achamy.html

12: FREQUENCY MUTATIONS

JOHN CHOWNING: AUDIO HACKER, SPATIALIZATION, DOPPLER SHIFTS, AND VIBRATO

THE BIRTH OF FM SYNTHESIS, FM MUTATIONS, TURENAS, STRIA AND THE GOLDEN MEAN

John Chowning, composer, in discussion with the author, December 19, 2023.

Center for Computer Research in Music and Acoustics. "Brief History.." https://ccrma.stanford.edu/~aj/archives/docs/all/646.pdf

Chowning, John. "John Chowning" https://ccrma.stanford.edu/people/john-chowning

Chowning, John. "The Synthesis of Complex Audio Spectra by Means of Frequency Modulation." https://ccrma.stanford.edu/sites/default/files/user/jc/fm_synthesis_paper.pdf

Chowning, John. "Fifty Years of Computer Music: Ideas of the Past Speak to the Future." https://ccrma.stanford.edu/sites/default/files/user/jc/SpringerPub3_0.pdf

Engineering and Technology History Wiki. "John Chowning" https://ethw.org/John_Chowning

Kirn, Peter. "Jean-Claude Risset" https://soundart.zone/jean-claude-risset-mutations-1969/

McGee, Ryan. "John Chowning: Overview, Techniques, and Compositions." https://lifeorange.com/writing/ChowningAnalysis_McGee.pdf

Rissett, Jean-Claude. *Music from Computer*. REGRM. 2014.

Rissett, Jean-Claude. *Computer Music Experiments 1964-* Computer Music Journal, Vol. 9, No. 1 (Spring, 1985), pp. 11-18

Roads, C. "John Chowning on Composition." https://ccrma.stanford.edu/~aj/archives/docs/all/143.pdf

Shirriff, Ken. "Reverse-engineering the Yamaha DX7 Synthesizer's Sound Chip from Die Photos." https://www.righto.com/2021/11/reverse-engineering-yamaha-dx7.html#fn:lavengood

Terman, Frederick E. *Radio Engineering*. New York, NY. McGraw-Hill, 1947.

Tsividis, Yannis. "Edwin Armstrong: Pioneer of the Airwaves." https://www.magazine.columbia.edu/article/edwin-armstrong-pioneer-airwaves

Zattra, Laura. "The Assembling of Stria: A Philological Investigation." http://digital.music.cornell.edu/center/wp-content/uploads/2020/01/40072593.pdf

13: SPEECH SONGS

THE MUSICAL POETICS OF PIERRE BOULEZ, SOUND WORD SYNTHESIS, AMBASSADOR OF THE AVANTGARDE

Benjamin, George. "George Benjamin on Pierre Boulez: 'He was simply a poet.'" https://www.theguardian.com/music/2015/mar/20/george-benjamin-in-praise-of-pierre-boulez-at-90

Boulez, Pierre. *Orientations: Collected Writings*. Cambridge, MA.: Harvard University Press, 1986.

Glock, William. *Notes in Advance: An Autobiography in Muisc*. Oxford, UK.: Oxford University Press. 1991.

Griffiths, Paul. "Pierre Boulez, Composer and Conductor Who Pushed Modernism's Boundaries, Dies at 90." https://www.nytimes.com/2016/01/07/arts/music/pierre-boulez-french-composer-dies-90.html

Jameux, Dominique. *Pierre Boulez*. London, UK.: Faber & Faber, 1991.

Peyser, Joan. *To Boulez and Beyond: Music in Europe Since the Rite of Spring*. Lanham, MD.: Scarecrow Press, 2008

Ross, Alex. "The Godfather." https://www.newyorker.com/magazine/2000/04/10/the-godfather

Shipwreck Library. "Joyce Music – Boulez: Répons" https://shipwrecklibrary.com/joyce/joyce-music-boulez-repons/

Sitsky, Larry, ed. *Music of the 20th Century Avant-Garde: A Biocritical Sourcebook*. Westport, CT.: Greenwood Press, 2002.

IRCAM, INTERCONTEMPORARY UNDERGROUND MUSIC, MAX/MSP

Georgina, Born. *Rationalizing culture: IRCAM, Boulez, and the Institutionalization of the Musical Avantgarde*. Berkeley, CA.: University of California Press, 1995.

IRCAM. "How Well Do You Know Espro?" https://manifeste.ircam.fr/en/article/detail/connaissez-vous-lespace-de-projection/

Krämer, Reiner. "X: An Analytical Approach to John Chowning's Phoné."
https://ccrma.stanford.edu/sites/default/files/user/jc/phone_kraemer_analysis_0.pdf

McHugh, Gene. "Don Slepian's 'Sunflower Geranium'" https://rhizome.org/editorial/2013/aug/12/don-slepians/

National Public Radio. "IRCAM: The Quiet House Of Sound" https://www.npr.org/templates/story/story.php?storyId=97002999#:~:text=IRCAM%20was%20created%20more%20than,20th%20century's%20pre%2Deminent%20composers.

Smith, Richard Langham, Potter, Caroline, ed. *French Music Since Berlioz*. Burlington, VT. Ashgate Publisher, Inc., 2006.

Tingen, Paul. "IRCAM: Institute For Research & Co-ordination in Acoustics & Music." https://www.soundonsound.com/people/ircam-institute-research-co-ordination-acoustics-music

CHANT, JONATHAN HARVEY'S RITUAL MELODIES, SPEAKINGS

Anderson, Julie. "Jonathan Harvey Dies Aged 73." https://www.takte-online.de/en/portrait/article/artikel/ircam-und-kathedral-chor-zum-tode-jonathan-harveys/index.htm

Bresson, Jean, Agon, Carlos. "Temporal Control over Sound Synthesis Processes." Sound and Music Computing (SMC'06), 2006, Marseille, France.

Chamorro, Gabriel José Bolaños Chamorro. "An Analysis of Jonathan Harvey's Speakings For Orchestra and Electronics." Ricercare No. 13, 2020.

Faber Music. "Jonathan Harvey's masterpiece trilogy at Edinburgh International Festival." https://www.fabermusic.com/news/jonathan-harveys masterpiece-trilogy-at-edinburgh-international-festival-252

Harvey, Jonathan. "Inner Light 1 (1973)" https://www.wisemusicclassical.com/work/7644/Inner-Light-1--Jonathan-Harvey/

Harvey, Jonathan. "Ritual Melodies." https://www.fabermusic.com/music/ritual-melodies-1504

Harvey, Jonathan. "Sprechgesang." https://www.fabermusic.com/music/sprechgesang-3850

Harvey, Jonathan. "Speakings." https://www.fabermusic.com/music/speakings-5282

Harvey, Jonathan, Lorrain, Denis, Barrière, Jean-Baptiste, Haynes, Stanley. "Notes on the Realization of 'Bhakti." Computer Music Journal, Vol. 8, No. 3 (Autumn, 1984), pp. 74-78 (5 pages)

Holmes, Thom. *Electronic and Experimental Music: Technology, Music, and Culture*. New York, NY.: Routledge, 2020.

Manning, Peter. *Electronic and Computer Music*. Oxford, UK.: Clarendon Press, 1993.

Rodet, Xavier, Potard, Yves, Barriere, Jean-Baptiste. "The CHANT Project: From the Synthesis of the Singing Voice to Synthesis in General." Computer Music Journal, Vol. 8, No. 3 (Autumn, 1984), pp. 15-31)

Service, Tom. "A Guide to Jonathan Harvey's Music." https://www.theguardian.com/music/tomserviceblog/2012/sep/17/jonathan-harvey-contemporary-music-guide

A FINAL WORD

Veltman, Chloe. "Send in the clones: Using artificial intelligence to digitally replicate human voices." https://www.npr.org/2022/01/17/1073031858/artificial-intelligence-voice-cloning Carter, Evans and

Novak, Analisa. "Scammers use AI to mimic voices of loved ones in distress." https://www.cbsnews.com/news/scammers-ai-mimic-voices-loved-ones-in-distress/

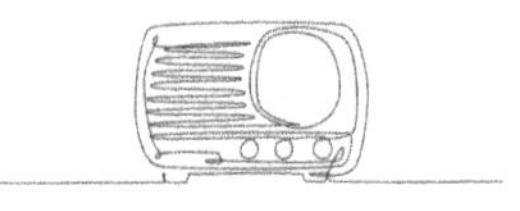

Acknowledgments

Radio and books are not made alone. They come out of the aether, and are accelerated along with the help of family, friends and colleagues. The following people, programs, stations, and shows all opened doors for me that might otherwise have remained closed.

Big ups to Cincinnati's WAIF and the programs and hosts: Art Damage, Alien Transmissions, Alien Soundtracks, Weird Trips, On the Way to the Peak of Normal, Trash Flow Radio, Hometown Hi-Fi, Uncle Dave Lewis, Jeremy Lesniak, Ron Orovitz, Andrew Hissett, Craig Kelley, John Cadwallader, AD8DM, and Ken Katkin. My time testing the waters of pirate radio on Anti-Watt prepared me for community radio and I am much obliged to Chris Monahan, Ben Hoffman, and Max Kruger.

I have had many stalwart companions for my many hours on the air at WAIF but I am indebted especially to Douglas Cholmondeley, "Monster" Syd Keeney and my late uncle, Daniel Moore. I don't know you all and I hardly know who any of you are, but I appreciate everyone who tuned in to my many hours of transmissions at WAIF.

A chance meeting with Brent Shields, KK4HMR led me to get my amateur radio license and join the Oh-Ky-In Amateur Radio Society. In that club I met my friend Robert Gulley, K4PKM, who has been a great elmer. His encouragement and enthusiasm for the initial articles that became the seed for this book gave me the impetus to keep writing.

DJ Frederick Moe, K1MOE, helped me realize a dream when he brought me aboard Skybird Radio and Imaginary Stations on the shortwaves. "One Deck" Pete Polyank, aka Madtone, M7MTE, a fellow trawler of the megahertz, has brightened my days by sharing in all of these high frequency high jinks.

To Jeff Braun, KF8PO, and Tom Pardi, NK8I, thanks for being net controls for the Big Event on Wednesday nights. To Dan Gettelfinger, KE8AWT, fellow WAIF junkie and radio fiend, I appreciate our time together playing radio and operating on Winter Field Day and Summer Field Day.

Many parts of this book originally appeared in different forms in the Q-Fiver, the newsletter of the Oh-Ky-In Amateur Radio Society. Thanks to the many members of the club for shared camaraderie, fun on the air, and all your combined hours of teaching and learning.

Thanks to the musicians, composers and sound designers, Dac Crowell, Matt Frantz, Hainbach and Erik T. Lawson. You served as ideal readers in my mind as I wrote these words. To all the members of Neato Torpedo, with special thanks to George Laub for asking me to join the band, and to Dave Schwinn for putting up with all my wrong notes. My gratitude goes out to Christian Hartman for his support, and Sean Morrissey for loaning me his Roland Juno-6. To all the organizers, bands and people who came out to gigs within the Cincinnati experimental music scene, I'm grateful for that time making collective noise.

To all my family and friends who have given me their counsel and heartened me along as I've traversed this twisting labyrinth of life, I am grateful. Without all your wisdom, I would surely be stuck in another dead end. Thanks for helping me see when I need to turn around.

To the memory of my mother, Julie Anne Moore, who encouraged me as a reader and a writer.

Thank you to the composers John Chowning, Paul Lansky, and Don Slepian for taking the time to speak and write to me about your work and lives.

Thanks to the Cincinnati and Hamilton County Public Library for providing all the new music and classical electronic albums that formed the inspiration for this book and to all the people who wrote the liner notes that I obsessed over.

Thanks to all the writers, journalists, researchers, documentary and video makers, radio show hosts, interviewers and interviewees whose work and information I was able to draw on, to write and synthesize this book.

Ryan Pinkard your deft hand as an editor has refined this manuscript, cutting the dross and bringing out whatever gold it may have, and made the work of cutting rather painless. Thanks for putting your shine on these words.

Colin Steven, thank you for cracking the door wide open, taking this book on and bringing it to readers.

Finally, thanks to everyone who bought the book on pre-order: Matthew Bielewicz, Urcun Bolkan, Nathanael Bonnell, Robert Borg, Jeffrey Braun, Erik Brueggemann, Anthony Burnetti, John Cadwallader, Audrey Cobb, Joy Cobb, Pam Cobb, Ron Cobb, Rae Cundiff, David Dalby, Gudmundur Erlingsson, Pedro Figueiredo, Tim Forrester, Matt Frantz, Oli Freke, Richard Frohman, Shaun Gannon, Stefan Goetsch, Gary Grant, Robert Gulley, Alan Holding, Anna Dechristopher Hopkins, Ken Katkin, Michael Klus, Ryan LaLiberty, Mark Lamb, Emilien Lesage, Fosco Lucarelli, Carol McBride, Adam Martin, Connie Moore, Jason Moore, Jonathan Moore, Joseph Moore, Victor Moore, Chris Monahan, Theodore Morris, Joerg Mueller-Kindt, Kent Mulcahy, Ron Orovitz, Stephen Page, Paul Palinkas, Cathleen Peters, Philip Rae, Wolfgang Schell, Steve Schenk, Jon Shute, Andrew Silagy, Pablo Smet, Phil Stuart, Liam Templeton, Mark Vaughan, Charity Ward.

Any errors or mistakes are all my own.